A-3-74

Introduction to
Computer Science Mathematics

Robert V. Jamison

Northrop Institute of Technology

Introduction to
Computer Science Mathematics

McGraw-Hill Book Company

New York Kuala Lumpur Panama
St. Louis London Rio de Janeiro
San Francisco Mexico Singapore
Düsseldorf Montreal Sydney
Johannesburg New Delhi Toronto

Library of Congress Cataloging in Publication Data
Jamison, Robert V.
   Introduction to computer science mathematics.
   1. Mathematics--1961-   I. Title.
QA39.2.J35   510   70-39901
ISBN 0-07-032276-7

INTRODUCTION TO COMPUTER SCIENCE MATHEMATICS

1 2 3 4 5 6 7 8 9 0 K P K P 7 9 8 7 6 5 4 3

*The editors for this book were Robert Flowers and Cath-
erine Kerr, the designer was Marsha Cohen, and its
production was supervised by James E. Lee. It was set
in Palatino by Progressive Typographers. It was printed
and bound by Kingsport Press Inc.*

Contents

1781280

Preface

This book is designed to present those special topics which are generally considered fundamental to the fields of data processing, computer programming, and computer design. Its level is that of a first course with one year of algebra considered prerequisite. The material is suitable for all students who will do any kind of computing, whether they will ultimately specialize in business or scientific and engineering applications or whether their interest will lie in strict computer science and computer design applications.

Finite mathematics concepts are covered in two chapters on the decimal number system and on other number systems, including the binary system. A comprehensive study of arithmetic in all these number systems is presented.

The function concept is used throughout. Linear, quadratic, polynomial, exponential, and logarithmic functions and equations are included. Additionally, they are linked together through the general topic of curvefitting. Numerical and useful semigraphical methods are explained and used extensively. Particular attention is paid to the linear function by means of such topics as linear equations, simultaneous linear equations, linear inequalities, linear programming, and matrix methods. Set theory is used throughout rather than being presented as a separate topic. Further topics in algebra should be studied by means of a full course in college algebra.

Matrices and determinants and their applications are discussed in a separate chapter. Included are applications of the method of elementary row operations to the finding of matrix inverses and to the solution of simultaneous linear equations.

Symbolic programming languages are discussed. A complete presentation is made of a somewhat shortened version of the Fortran IV language. Fortran was chosen as the language most suitable for the solution of the types of problems presented in this book. Basic Fortran IV is presented, including such topics as real and integer constants, arithmetic statements, input and output statements with Format, and control statements. Subprogramming is not discussed. Complete executable Fortran programs can be written following the study of this material. The topics of program planning and flowcharting are presented, and Fortran programs are used to solve problems throughout the subsequent portion of the book. Once again, those who will specialize in programming to solve advanced problems in the scientific and engineering fields can then undertake a full course in Fortran programming and with this background should succeed with great ease.

The rather extended chapter on Boolean algebra proceeds from the basic physical properties of the simple switching circuit to a full study of the theory and further applications of Boolean algebra. Such topics as Boolean identities, truth tables, and normal forms for Boolean expressions are covered. Karnaugh maps are used extensively as the main method for simplifying Boolean expressions. Throughout, logic blocks are used to design logic circuits that implement Boolean functions. Functionally complete logic blocks such as NAND and NOR are explained and employed. An insight into the design of the arithmetic unit of a digital computer is provided by the section on half- and full-adders.

Problem sets are interwoven throughout the chapters immediately following the presentation of the topics to which they apply. Answers to selected exercises appear at the end of the book.

The author is indebted to Robert D. Chenoweth, County College of Morris, Robert L. Monsees, Meramec Community College, and Forrest L. Barker, Los Angeles City College, for their critical reviews of the original manuscript and their valuable suggestions for its improvement.

Robert V. Jamison

Introduction to
Computer Science Mathematics

1

The Decimal Number System

1.1 INTRODUCTORY REMARKS

The first chapter of this book is about decimal numbers because they are the objects that we write, manipulate, and combine in various ways to do all the computations that are required to solve problems. We do not intend to discuss the theoretical and abstract nature of numbers. Instead, we shall often assume that you are familiar with many of the properties of numbers simply because you have used them so often in presumably reasonable and useful applications. Sometimes, though, we shall present interesting properties of numbers in unusual ways in order to provoke some new thinking on your part. It is extremely important that we understand the properties of numbers and the rules of arithmetic before we undertake the study of algebra. After all, algebra is simply a generalization of the arithmetic of numbers in which nonnumeric symbols are often used to represent numbers.

No matter how delightfully descriptive (or mischievously deceptive) the language in which a problem in science, engineering, business, or otherwise is stated, the words must be translated into algebraic and other kinds of relations and equations among the numbers and symbols representing numbers. To make this possible, we shall always require that the problem be inherently amenable to mathematical solution; that is, it must be well posed. This translation must accurately restate the problem in mathematical form. It therefore seems fairly evident that careful reading and logical analysis of the problem as well as familiarity with applicable mathematical language are absolutely essential for making the correct mathematical interpretation.

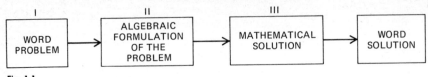

Fig. 1.1

With all the rhetoric then removed, the problem becomes one on which the principles and techniques of applied mathematics can be used, that is, the tools of analysis and computing you will learn here and in later courses in applied mathematics. The solution of the problem can then be translated back into the language in which it was originally stated.

The process just described can be graphically represented by the block diagram of Fig. 1.1. All problems from whatever source must be translated into the form shown in block II and solved by mathematical techniques indicated by block III without any consideration as to the original meaning of the numbers, symbols, and relations involved. In this sense, applied mathematics is abstract. Much of your work will be concerned with the achieving of the solution required in block III after the problem has been properly stated mathematically. Experience through much practice is the only reasonable way of acquiring the facility in thought and deed necessary to achieve proper and accurate results.

1.2 THE DECIMAL NUMBER SYSTEM

Any number is constructed from a (usually) small set of basic numbers called *digits*. As you recall, the digits of the decimal number system are the 10 integers 0, 1, 2, 3, 4, 5, 6, 7, 8, and 9. It is from the number of digits that the system takes its name. If we include the three symbols + (the plus sign), − (the minus sign), and . (the decimal point), we can construct every number of the decimal number system. Any number of the system is simply a sequence of these digits and symbols formed by placing them horizontally and in close proximity, that is, next to each other. Of course, a few rules must be followed. The minus sign, as you already know, need not occur at all and if it does, it must appear as the first symbol of the number and can occur only once in each number. The decimal point need not appear either but if it does, it can occur only once and can be in any position in the sequence. Thus, we can have the numbers −43.7000, 0.1324, 8475663, and −3.002.

Numbers whose first symbol is the minus sign are called *negative* numbers; for example, −3.56 and −4567723 are negative numbers. Those numbers that do not start with the minus sign are called *positive* numbers; for example, 0.0098 and 2345.7 are positive numbers. Sometimes the positiveness of a number is emphasized by starting the sequence with the symbol + (the plus sign). Thus, we sometimes write +0.0098 and +2398.7, but we shall continue to omit the plus sign before numbers if its use is merely to indicate that the number is positive.

Numbers that contain no decimal point are called *integers*. Thus, 0, −12, and 3753 are integers. Obviously, there are positive and negative integers. The positive integers are often used for counting. We shall later also use them as subscripts.

The sequence of symbols that occur before the decimal point is called the *integral* or whole part of the number. For example, 17.38 has 17 for its integral part; −896.7 has −896 for its integral part. The integral part of a number is an integer. The part of the number that starts with the decimal point and includes all the digits that follow the decimal point is called the *fractional* part of the number. 17.38 has .38 as its fractional part; 0.083 has .083 as its fractional part. Integers, like 255, of course, have no fractional part.

We shall see when we discuss in Chap. 4 the computer language called Fortran that much is made of the difference between numbers that do and do not contain a decimal point. In that language a number is called an integer if it does not contain a decimal point, just as we have defined it. It is called *real* if it does contain a decimal point. Adding a decimal point after the 7 in 17 changes the integer 17 to the real number 17. in that language.

Nothing has yet been said about how many digits a number may contain. None of the sample numbers we have shown so far have had very many. In your experience in computing with numbers of the decimal system, you undoubtedly preferred numbers that contained relatively few digits. Many desk calculators have room in their register for only 12 digits, and all digital computers, although they may allow many more digits than 12 in their registers, still definitely restrict the number of digits in all the numbers they operate on. Here is where the kind of numbers used in theoretical mathematics and those used in actual computing work differ. Let us first discuss numbers that are hypothetically allowed in mathematics.

Any quantity of digits, including sequences which go on indefinitely and never stop, is allowed in numbers. Thus, 37.828282 . . . could indicate that the number has infinite repetitions of the terminal sequence 82. For a number of this kind we shall prefer the notation 37.(82), in which the parentheses around 82 are used to indicate the infinite repetition of the terminal sequence 82. In the representation 37.828282 . . . , the three decimal points indicate that the pattern thus presumably set is to be repeated indefinitely. A number need not terminate (that is, have a last digit) nor repeat a terminal sequence at all. In this case, we would use the . . . notation. For example, 2.14935876 . . . could indicate a number in which no repeating pattern of digits ever occurs.

Now, a number which terminates or which has a terminal repeating sequence like 37.(82) is called a *rational* number. Any rational number can be represented exactly as the quotient of two integers; in fact, this property of rational numbers is often taken as their defining property. In Exercise 1.2 we shall present a method for finding these integers for a rational number presented in decimal form. It seems fairly obvious that all integers are rational numbers according to this last criterion. For example, the integer 17 can be written as 17/1, the ratio of two integers, and hence is a rational number.

A number which does not terminate and which never exhibits a repeating sequence no matter how far one continues its digits is called *irrational*. You recall that numbers which you have previously symbolized as $\sqrt{2}$, $\sqrt[4]{7}$, and π are irrational. They are thus represented by digit sequences that do not terminate and never have a repeating terminal sequence. An irrational number cannot be expressed as the quotient of two integers.

The entire collection of all rational and irrational numbers is called (in mathematics) the set of *real* numbers. In Chap. 6 we shall define a new kind of number called *complex*. We shall see that every real number is a complex number but that there are complex numbers which are not real.

As we mentioned above, in computing either by hand or by any kind of mechanical or electronic computer, all numbers must be represented by a sequence of digits that does terminate. When computing by hand, carrying too many digits in the numbers puts a strain on anyone's patience and time.

Suppose a calculator allows five digits, regardless of the position of the decimal point in the number, if it has one. Clearly, one cannot insert such irrational numbers as 38.25827546 . . . , such rational numbers as 7.(3), or even such integers as 84670995 into the calculator. To use such numbers in this machine, one must first cut them off in some way to a maximum of five digits. We shall return to this project and discuss the notion of significant figures in Sec. 1.3.

We must now turn to the notion of the *value* of the numbers we have constructed. Long experience with numbers has no doubt revealed to you that somehow, even though the numbers 861 and 168 are composed of the same set of integers, the digits are in different orders and the numbers have different "values." This is implied by the very way we read the numbers. When we read the number 861, we do not just say "eight six one," we say "eight hundred sixty-one."

The simple but important concept of how the *position* of the digits in a number determines its value is explained below. Consider the following diagram:

4	3	2	1	0

The blocks indicate the positions the digits of the number will occupy. They are numbered as shown starting with 0 at the block farthest to the right and proceeding to the left. Associated with each digit in a number is a value called its *positional value*. It is determined in the following way: The positional value of a digit in position 0 (called the *units position*) is $10^0 = 1$; the positional value of a digit in position 1 (called the *tens position*) is $10^1 = 10$; in a similar way, in position 2 (called the *hundreds position*) a digit has positional value $10^2 = 100$; and in position 3 (the *thousands position*) a digit has positional value $10^3 = 1000$. The same process continues as indicated until the first digit of the number has been reached and assigned its positional value. We finally arrive at the following rule: The *value* of a number is the sum of the product of each digit of the number and its respective positional value. Thus,

3	2	1	0
8	6	4	4

has value $8 \times 10^3 + 6 \times 10^2 + 4 \times 10^1 + 4 \times 10^0 = 8000 + 600 + 40 + 4$, which we read "eight thousand six hundred forty-four." This is the value of the number that we write, using only the digits underlined, as 8644.

The method just used gives the value of numbers which either are integers or have a decimal point immediately after their last digit. Let us now see how we find the value of numbers that have zero integral part. Recall that $10^{-1} = \dfrac{1}{10} = 0.1$, $10^{-2} = \dfrac{1}{100} = 0.01$, $10^{-3} = \dfrac{1}{1000} = 0.001$, and so forth, so that a nonzero number with zero integral part can have its digits placed in the following diagram:

−1	−2	−3	−4	−5

Now, the positional value of a digit in position −1 (called the *tenths position*) is $10^{-1} = 0.1$; the positional value of a digit in position −2 (called the *hundredths position*) is $10^{-2} = 0.01$; similarly, a digit in position −3 (called the *thousandths position*) has positional value $10^{-3} = 0.001$; we can continue in the same way until we reach the last digit of the number and its positional value is assigned. Thus,

−1	−2	−3	−4	−5
3	0	7	5	9

has value $3 \times 10^{-1} + 0 \times 10^{-2} + 7 \times 10^{-3} + 5 \times 10^{-4} + 9 \times 10^{-5} = 0.3 + 0.00 + 0.007 + 0.0005 + 0.00009$, which we write, using the digits underlined, as

0.30759. The number in this example has an interior zero digit. If we pose a problem in reverse — write a number with value $3 \times 10^{-1} + 8 \times 10^{-3} + 9 \times 10^{-5}$ — we must insert the digit zero in the appropriate positions in the number so that the proper value will be obtained by using the method described above. The number is 0.30809.

More generally a number can have both integral and fractional parts. For example,

2	1	0	−1	−2	−3
8	6	6	0	5	9

has value $8 \times 10^2 + 6 \times 10^1 + 6 \times 10^0 + 0 \times 10^{-1} + 5 \times 10^{-2} + 9 \times 10^{-3} = 800 + 60 + 6 + 0.0 + 0.05 + 0.009$, which we write 866.059.

All this probably seems self-evident to you, but only because your previous use of numbers has impressed the notion of position and its connection with value deeply in your consciousness. Actually the notion of position is not at all obvious, and the positional notation which is so natural to us today has evolved after years of thought and refinement on the part of mathematicians.

EXERCISES

1.1 Represent each of the following rational numbers in the form of infinitely repeating decimals. Use the abbreviated notation described in the text. Example: $2./3. = 0.(6)$.

(a) 16./11.
(b) 53./15.
(c) 4./7.
(d) −17./23.
(e) 35./4.
(f) 11./9.

1.2 The repeating decimal 1.(63) can be represented as the ratio of two integers by following this procedure: Let $p = 1.63636363 \ldots$; then $100p = 163.63636363 \ldots$. By subtraction, $99p = 162$. Therefore, $p = 162/99 = 18/11$. Use this procedure to represent the following repeating decimals as the ratio of two integers:

(a) 0.(23)
(b) 1.6(4)
(c) −0.0(325)
(d) 0.0000(87)

(e) 4.(6034)

(f) 0.45(7)

(g) 12.(5)

(h) −0.36(72)

(i) 0.00(707)

1.3 The notation $[x]$ is often used to denote the greatest integer contained in the real number x. Thus, $[4.32] = 4$, $[0.37] = 0$, and $[72.999] = 72$. Show that if x is not negative, $[x]$ is the integral part of x. What is $[-13.72]$? $[-0.0678]$?

1.4 Use the power-of-10 method to write the value of the following numbers. Example: $87.3 = 8 \times 10^1 + 7 \times 10^0 + 3 \times 10^{-1}$.

(a) 67995

(b) 0.009506

(c) 4.586

(d) 736.947

(e) 50.009

(f) 12.12125

1.5 Write the number whose value is

(a) $6 \times 10^3 + 8 \times 10^2 + 0 \times 10^1 + 8 \times 10^0$

(b) $9 \times 10^{-1} + 8 \times 10^{-4}$

(c) $7 \times 10^1 + 5 \times 10^0 + 4 \times 10^{-2}$

(d) $8 \times 10^{-3} + 4 \times 10^{-4} + 3 \times 10^{-5} + 7 \times 10^{-7}$

(e) $4 \times 10^2 + 6 \times 10^0 + 7 \times 10^{-1} + 7 \times 10^{-2} + 8 \times 10^{-3}$

1.3 STANDARD FORM. SIGNIFICANT FIGURES. ROUNDING OFF

Often it is desirable to write a real number in what is called *standard form*. This has to do with the position of the decimal point in the number. If the decimal point immediately precedes the first nonzero digit of the number, we say that the number is written in standard form. Thus, 0.30472 is in standard form but neither 17.30 nor 0.000976 is. In order to place a number in this form, clearly a way must be found to move the decimal point in a number left or right as required but without changing the value of the number.

Suppose one has a number; it should be fairly obvious that if one allows the decimal point to float up and then to the right over succeeding digits, each time it passes a digit the number is multiplied by 10. For example, 86.4632 results in 86463.2, which is 10^3 or 1000 times the original number. To maintain the value of the original number without moving the decimal point back, the resultant number must be multiplied by 10^{-3}. This can be done by writing 86463.2×10^{-3}.

Again, if the decimal point floats to the left over five digits, the number is actually divided by 10^5, and to compensate, the resultant number must be multiplied by 10^5. Thus, ⌢00098.43 results in 0.0009843 and $98.43 = 0.0009843 \times 10^5$. Note the zeros that were inserted because of positional considerations.

Now, all this discussion leads to the following rule for placing a number in standard form: Move the decimal point until it is immediately before the first nonzero digit in the number. If the decimal point has been moved left, multiply the result by 10 raised to the power equal to the number of places moved. If the decimal point has been moved right, multiply the result by 10 raised to the power equal to the negative of the number of places moved. Thus, $76.73125 = 0.7673125 \times 10^2$ and $0.000001025 = 0.1025 \times 10^{-5}$.

Once a number has been placed in standard form, it can be stored (that is, placed) in the register of a calculator or computer in the following two-block form:

| 7 | 6 | 7 | 3 | 1 | 2 | 5 |

| + | 0 | 2 |

and

| | 1 | 0 | 2 | 5 | 0 | 0 |

| − | 0 | 5 |

The number in the first block in each case is called the *mantissa* of the number; the second number (with its sign) is called the *characteristic* of the number.

This mantissa and characteristic method (and standard form) enables one to write very large or very small numbers without using all the zeros that might be necessary in ordinary decimal form. For instance, 765540000000000. can be written as 0.76554×10^{15} and 0.0000000000000013401 can be written as 0.13401×10^{-14}.

We shall later see that in the computer language called Fortran a special code is used to indicate a number written in the form of mantissa and characteristic. There we shall write 0.86732×10^4 as .86732E+04 and 0.1025×10^{-8} as .1025E − 08. Clearly, the combination of symbols E+04 is replacing the multiplier $\times 10^4$, and E − 08 is replacing $\times 10^{-8}$.

Up to now we have been writing numbers of all sorts and sizes without any information about their source—that is, without knowing or caring about what is sometimes called the *accuracy* of the number. When we write 17.38, there is absolutely no way of telling how "accurate" the number is by merely looking at it. However, if we knew that this number resulted from a rounding off of another number previously known exactly, then we would have some idea as to its accuracy. We shall discuss the procedure of rounding off in a moment, but in the absence of any prior knowledge of it, when we write 17.38 we mean 17.38 exactly and not approximately.

Here we must discuss a concept concerning numbers called *significant figures*, because of its current common usage. One must be cautioned, however, not to place too much emphasis on the usual meaning of the word "significant" as it is used here. The number of significant figures a number contains is simply the number of digits it contains, the number of digits being counted using the following rule: Count the number of digits in the number (no matter where the decimal point is) starting with the first nonzero digit. Thus, 86.472 has five significant figures, 0.003705 has four significant figures, and 14.3700 has six significant figures. Note that zeros occurring after the first nonzero digit are included in the count. Some special cases occur. For example, 837000. has three significant figures when one presumes that the three zeros are present merely for positional purposes; but if the number is known to be exact, then 837000. has six significant figures.

As indicated in the last section, numbers sometimes have so many significant figures that it is physically next to impossible to do all the operations of a set of calculations and throughout maintain all the figures the given numbers and the intermediate results contain. This is true not only when one is computing by hand with pencil and paper but also when one is computing with the aid of the fabulous computer. Here the notion of the approximate value of a number becomes important. You recall that the important irrational number symbolized by π is represented as a decimal number by an infinitely long, never-repeating sequence of digits (since it is irrational). The sequence starts with 3.14159 and continues indefinitely. In order to do any calculation involving π, it is fairly clear that one must approximate it (that is, replace it with a rational number by terminating its sequence of digits at some reasonable finite place). One could approximate π by writing $\pi \doteq 3.1$ (to two significant figures). Here we are using the new symbol $\doteq$ to replace the words "is approximately equal to." One could also write $\pi \doteq 3.14$, $\pi \doteq 3.141$, $\pi \doteq 3.1415$, or $\pi \doteq 3.14159$, depending on whether one wants three, four, five, or six significant figures in the approximation. In this context all these numbers are approximate in the sense that they are known to be approximations of the number π. But, obviously, if one were to see 3.14 in a completely unrelated context, one would not conclude that in that situation 3.14 would be approximating π. It could perfectly well be that 3.14 is an exact number, perhaps the decimal equivalent of the mixed fraction 3 $\frac{7}{50}$. It could just as easily have been the result of rounding off the exact number 3.144627 to three significant figures.

It is clear that all decimal numbers with a large, perhaps infinite, quantity of digits must be rounded off at some finite stage even if they are rational numbers.   Recall that $1/3 = 0.(3) = 0.333333$ . . . .   Thus, $1/3$ can be approximated by 0.3, 0.33, 0.333, 0.3333, etc., depending upon the number of significant figures one wants, or perhaps more often depending upon the capability of storage of the computer one is using.   Here the rounding off is achieved by simply cutting off the sequence at the selected place and completely omitting the remaining digits of the sequence, whatever they were.   In a case like this one, if the choice of how many 3s to carry in a certain calculation is completely up to you, we suggest that you carry as many significant figures as you physically can or as many as are feasible in the computing device you are using. One can round off a number that already has a terminal digit if one cares to. For example, 873.2 can be rounded off to 870. (and thus reduced from four to two significant figures) by omitting the 3.2 and supplying the 0 needed for positional accuracy.   It is hard to think of the circumstances which would cause one to do this, however.

There is a kind of rounding off called *symmetric*.   It is the method that you are perhaps familiar with even though it is not the method used in most computers; they usually automatically round off to a prescribed number of digits (depending upon the particular machine) by simply cutting off (truncating) the sequence of digits and replacing digits by zeros when necessary. Here is the rule for symmetric rounding: Starting with the first significant figure, count to the right the number of places equal to the number of significant figures desired in the result.   For example, 0.0083625 says that three significant figures are wanted.   Let us call the very next digit the *roundoff digit*. We then follow these steps:

1. If the roundoff digit is 4 or less, simply omit the roundoff digit and the remaining sequence of digits:

   $0.0083625 \doteq 0.00836$ (three significant figures)

   and

   $0.9563499 \doteq 0.9563$ (four significant figures)

2. If the roundoff digit is 6 or more, advance the digit just before it by 1 and omit the roundoff digit and the remaining sequence of digits:

   $7.83694 \doteq 7.84$ (three significant figures)

3. If the roundoff digit is 5 and at least one nonzero digit follows it eventually, advance the digit just before it by 1 and omit the roundoff digit and the remaining sequence of digits:

$$87.4650004 \doteq 87.47 \qquad \text{(four significant figures)}$$

4. If the roundoff digit is exactly 5 (with no nonzero digit ever appearing after it), advance the digit just before the 5 by 1 if it is odd, but do not advance it if it is even, and once again omit the roundoff digit and the remaining sequence of digits:

$$862.5 \doteq 862 \qquad \text{(three significant figures)}$$

but

$$0.0093555 \doteq 0.009356 \qquad \text{(four significant figures)}$$

Note again that most digital computers do not use built-in symmetric rounding but instead automatically round off by the simple process of truncating, that is, cutting off the sequence of digits. It is possible, of course, for a program to be written and stored in the computer which will allow one the option of symmetric rounding.

EXERCISES

1.6 Express each of the following numbers in standard form:

(a) 4862.47
(b) -0.007053
(c) 337.0×10^6
(d) $42832. \times 10^{-7}$
(e) 93470000000000000000.
(f) 0.000000000045688
(g) -4785.003

1.7 Find the mantissa and the characteristic of each of the following numbers:

(a) 869.37
(b) 4283000.
(c) 625×10^{16}
(d) 6
(e) 497×10^{-3}
(f) 0.00098907

1.8 Express each of the numbers of Exercise 1.6 using the Fortran E notation.

1.9 State the number of significant figures in each of the following numbers:

(a) 17.320
(b) 0.008326
(c) −13.227
(d) 98300.7
(e) 93902
(f) 1.00467
(g) 100.001
(h) 1.1111
(i) 78

1.10 Express each of the following numbers to the given number of significant figures using symmetric rounding:

(a) 17.325 (4)
(b) 887.12 (1)
(c) 0.003295 (2)
(d) 637.5 (3)
(e) 985435 (5), (4), (3), (2)
(f) 14.5678 (2), (3)
(g) 789.07 (4), (2)

1.11 Express each of the numbers of Exercise 1.10 to the given number of significant figures using *truncation*.

1.4 OPERATIONS

We shall assume that you are perfectly capable of doing the arithmetic involved when you perform the operations of addition, subtraction, multiplication, and division of real numbers. Some confusion seems to exist, however, about the accuracy of the result when numbers are combined using these operations.

Suppose someone asked you to add 8.37, 4.2, and 863. You undoubtedly would place the numbers in a column with the decimals lined up and add as usual:

```
  8.37
  4.2
863.
------
875.57
```

The result would be 875.57. Since the three numbers are presumed to be exact, the result is exactly 875.57. Then, if for some unknown reason the result is desired to only two significant figures, we replace 875.57 by 870. (using truncation) or by 880. (using symmetric rounding). We never round off the original numbers to two significant figures before adding.

Similarly, when 7.5 is multiplied by 3.7, the result is 27.75. We do not presume that the numbers 7.5 and 3.7 are somehow approximate merely because they contain only two significant figures. The result is exactly 27.75.

There are, of course, instances where roundoff of the result of a calculation must occur. Suppose we are using a calculator which has three registers for the use of the operation of multiplication—the multiplicand, multiplier, and product registers. Suppose each of these registers is made up of two parts, each allowing three symbols. The first contains the mantissa of three digits with the decimal point assumed as preceding the first digit. The second part contains the characteristic, the power of 10 by which the number in the first section is to be multiplied. For example,

| 1 | 2 | 3 | | − | 0 | 2 |

is the way a register looks when it contains the number 0.123×10^{-2}.

Suppose that the multiplicand and multiplier registers contain, respectively,

| 1 | 2 | 3 | | − | 0 | 2 | and | 4 | 6 | 9 | | + | 0 | 5 |

(representing 0.123×10^{-2} and 0.469×10^{5}). Now let us suppose that the computer takes these two numbers to some internal calculating area where there is plenty of room to perform the multiplication carrying all figures, rounds off the product to three significant figures (by cutting off the proper terminal digits), and returns to the product register the result in standard form. Then the following occurs internally:

$$
\begin{array}{r}
0.123 \times 10^{-2} \\
0.469 \times 10^{5} \\
\hline
1107 \\
738 \\
492 \\
\hline
\end{array}
$$
$0.057687 \times 10^{3} = 57.687 = 0.57687 \times 10^{2}$

Notice that the first product is found by multiplying as usual the numbers in the first parts of the two registers and supplying the decimal point in the result as usual in arithmetic. The numbers that appear in the second part of the two registers are *added*, that is, $(-2) + (5) = 3$, and the result is the power of 10 by which the first product must be multiplied. The reason the numbers -2 and 5 are added is that we are actually multiplying 10^{-2} by 10^5 to get 10^3 and the laws of exponents require us to add the exponents.

As shown above, the next step is to place the result 0.057687×10^3 in standard form. Then, since there is no symmetric rounding, the following will appear in the product register:

5	7	6		+	0	2

The 87 on the end of 57.687 has been truncated.

The two registers could contain two numbers that are to be added, for example,

8	6	3		+	0	2		and		1	2	5		+	0	1

representing 0.863×10^2 and 0.125×10^1.

Then, if the add button is pushed, the following calculation could ensue internally:

$$\begin{array}{r} 86.3 \times 10^0 \\ \underline{1.25 \times 10^0} \\ 87.55 \times 10^0 \end{array}$$

Note that now the characteristics are treated in a different manner. First, the characteristics must be the *same* before the numbers are added. In the example just above, we have made the characteristics both zero by writing 0.863×10^2 as 86.3×10^0 and 0.125×10^1 as 1.25×10^0. We then add the new mantissas and multiply the result by this common power of 10. We could just as well have written 0.863×10^2 as 0.0863×10^3 and 0.125×10^1 as 0.00125×10^3, and then have added in this way:

$$\begin{array}{l} 0.0863 \times 10^3 \\ \underline{0.00125 \times 10^3} \\ 0.08755 \times 10^3 = 87.55 \text{ (as before)} \end{array}$$

In any case, the following will appear in the sum register:

8	7	5		+	0	2

Obviously, the numbers that we have shown appearing as if by magic in the various registers of the computer could very easily have been themselves the rounded-off results of previous calculations. In a sequence of calculations where roundoff occurs at the end of each operation, the final result will have accumulated roundoff errors. This sort of buildup is an inherent property of computers and is one of the most troublesome problems that one has to consider when evaluating the accuracy of the result of a set of calculations. It has simply to do with the fact that the various registers of the computer are capable of storing numbers that are finite in length. If one is computing by hand, then of course one can maintain many more figures throughout and thus protect the accuracy of earlier results. In this case one should keep at least one (and preferably more than one) figure in excess of the number of significant figures one desires in the result. If possible, never round off intermediate results. In any case, never round off to the final number of figures except in the very last result. Note that in some computers it is possible to achieve much this same effect of keeping more figures throughout a set of calculations by using the computer's capability called *extended precision*. When one can use this, even when the numbers (data) inserted into the computer for it to perform calculation on are of few significant figures, the computer automatically maintains quite a few more figures continually and thus helps alleviate the problem of roundoff error.

Here is an example which compares the calculation by hand with no rounding except in the final result and by use of a computer with rounding-off characteristics we have just discussed (not extended precision).

Example. Compute $(8.2 \times 17.9 + 2.87) \times 3.7$. Presume the numbers are exact and round off the result to three significant figures. Method 1: Exact calculation carrying all figures throughout.

```
        8.2          146.78             149.65
  ×    17.9      +    2.87       ×         3.7
       738          149.65             104755
       574                             44895
        82                            553.705
     146.78
```

Result: 553.

Method 2: Use the computer method with roundoff (truncation) as required at the end of each operation.

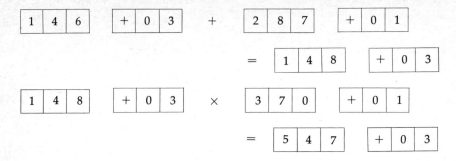

We leave it to you to check the intermediate results shown. In any case, the final result of the computer method is 547.

Notice the "error" introduced in the final result by the accumulation of roundoff error even though all the given numbers, the data, fit exactly into the computer registers with no initial roundoff required to store them.

The operation of division will often lead to infinite repeating decimals which must necessarily be rounded at some finite place. For example, $8./7. = 1.(142857) \doteq 1.143$ to four significant figures (in this case using symmetric rounding). We repeat that if the final result of a sequence of operations is to be rounded to a certain number of significant figures, one should maintain one or two more significant figures than this number in all the intermediate calculations.

EXERCISES

1.12 Assume that each of the numbers given below is exact. Perform the indicated operations by hand carrying all figures. Round the final result to three significant figures using truncation (cutting off).

(a) 17.3×0.82

(b) 149.3×7.1

(c) $8.267 + 5.92 + 13.2 + 17. + 647.8957$

(d) $(13.67 \times 12.5) + 72.8$

(e) $138.75 - 67.983$

(f) $0.678 \times 4.3 \times 0.06$

(g) $\dfrac{987.}{3} + 6.389$

(h) $(23.45 - 9.56) \times 12.6$

(i) $\dfrac{89.5 + 46.7}{13.} - 9.57$

1.13 Assume that a computer has a four-digit register for the storage of numbers. For example, $\boxed{8\,|\,1\,|\,4\,|\,7}\;\boxed{+\,|\,0\,|\,0\,|\,1}$ represents $0.8147 \times 10^1 = 8.147$. Assuming rounding by cutting off (truncation) at the end of each operation, perform the following sequence of operations. Place the result in the register form shown.

(a) $0.6428 \times 10^5 + 0.2375 \times 10^2 - 575.0$
(b) $(8.32 \times 17.76) + 2.376$
(c) $8.27/7. - 0.003728 \times 10^2$
(d) $(6.28) \times (72.9)/8.$
(e) $(8.37 + 0.6578 + 17.379) \times 4.72$
(f) $2.632 \times 2.632 - 4. \times 1.873 \times 6.03$
(g) $(56.78 + 68.79 + 45.68 - 34.77)/0.4$
(h) $5.678 \times 34.5 + 3.45 \times 32. + 4. \times 5.679 + 5.6 \times 34.6$

1.14 Perform the sequences of operations of Exercise 1.13
(a) assuming that the computer has a *three*-digit register for the storage of numbers—for example, $\boxed{7\,|\,8\,|\,9}\;\boxed{-\,|\,0\,|\,4}$ represents $0.789 \times 10^{-4} = 0.0000789$;
(b) assuming that the computer has a *five*-digit register for the storage of numbers—for example, $\boxed{6\,|\,7\,|\,8\,|\,9\,|\,8}\;\boxed{+\,|\,2\,|\,3}$ represents the number 0.67898×10^{23}.

1.15 For each of the parts of Exercise 1.13, compute exactly each operation using *all* figures throughout. Then find the error in each answer of Exercises 1.13 and 1.14.

1.5 CONCLUDING REMARKS

In the next chapter we shall introduce real-number systems with bases other than 10. The four arithmetic operations in these systems will prove new, fascinating, and, interestingly enough, very practical. Several of these systems, namely the binary, octal, and hexadecimal, are particularly important in computer applications.

2

Nondecimal Number Systems

2.1 INTRODUCTORY REMARKS

Any positive integer greater than 1 can be used as the base of a number system. The digits of a number system using a base b which is 10 or less are 0, 1, 2, 3, . . . , $b-1$. For example, the digits of number base 8 are 0, 1, 2, 3, 4, 5, 6, and 7. We shall see in a moment how we handle the situation of the digits of a number system with base more than 10. In any case, the numbers of the system are then constructed in exactly the same way we used for the decimal system except, of course, that the digits used are restricted to those belonging to the system. Thus, 630.72 is a very respectable number in the base 8 number system. As before, 630 is the integral part of the number and .72 is its fractional part. And, even though the system is no longer decimal, the "decimal point" which, in a sense, joins the two parts of the number is used.

Systems with certain bases, because of their importance in computer applications, have been given specific names (like *decimal* for base 10). Some of these are the binary system (base 2), the ternary system (base 3), the octal system (base 8), and the hexadecimal system (base 16). Systems with bases other than these must be referred to by giving the base itself.

One notices immediately that if the number 630.72 is standing alone, out of any context which indicates the base being used, it would only be natural to assume that the number is decimal. So some way must be found to designate the base if it is necessary for clarity. The method we shall use here is to follow the number by a subscript indicating the base; thus we would write $630.72_{(8)}$ to indicate that the number is in the octal system. Generally, if no subscript follows a number, we shall presume that the number is decimal.

The binary system, base 2, is of such paramount importance that much attention will be paid to it in the following discussions in this chapter and in later chapters when specific applications are explained. It has the neat property of requiring only two digits, 0 and 1. So, 110111011.1101 could be a number in the binary system.

Suppose, however, that one wants to use a base greater than 10, say 12. Then the digits we shall use will be 0, 1, 2, 3, 4, 5, 6, 7, 8, 9, A, and B. Notice that since 12 digits will be required and there are only 10 decimal digits, we have used the single symbol A to stand for the decimal number 10 and B to stand for the decimal 11. A39B1.8A is a number in this system. We shall not use numbers of base greater than 16, but, in general, the digits for a number of base greater than 10 could be 0, 1, 2, 3, 4, 5, 6, 7, 8, 9, A, B, C, D, E, F, G, ... In the hexadecimal system, base 16, the first 16 of these are used, so that A, B, C, D, E, and F stand for the decimal numbers 10, 11, 12, 13, 14, and 15, respectively, but are *digits* in the hexadecimal number system. Thus, 39F7E.4BA is a hexadecimal number.

2.2 THE DECIMAL VALUE OF NONDECIMAL NUMBERS

We now turn to the important notion of the *decimal* value of numbers in non-decimal systems. We shall consider first numbers with no fractional part, that is, whole numbers. Suppose we are using base 7 and have the number $3205_{(7)}$. The following shows the *positional* values we assign to each digit:

Number   3   2   0   5
Positional
   value 7^3 7^2 7^1 7^0

We notice, of course, the direct analogy with the positional values we discussed in Chap. 1 for decimal numbers. Here, however, the positional values are powers of 7 and not of 10. Now we come to the rule for calculating the *decimal* value of the number: The decimal value of a number is found by multiplying each of its digits by its respective positional value and adding these products. Here the multiplication and addition required are done using the ordinary *decimal* arithmetic. Thus, to find the decimal value of $3205_{(7)}$, we have

$$3 \times 7^3 + 2 \times 7^2 + 0 \times 7^1 + 5 \times 7^0 = 3(343) + 2(49) + 0 + 5(1)$$
$$= 1029 + 98 + 5$$
$$= 1132$$

That is, $3205_{(7)} = 1132_{(10)} = 1132$.

Example. The decimal value of $3E5B_{(16)}$ is found by multiplying each of its digits by its respective positional value and adding these products. Thus,

$$3E5B_{(16)} = \underline{3} \times 16^3 + \underline{E} \times 16^2 + \underline{5} \times 16^1 + \underline{B} \times 16^0$$
$$= 3(4096) + 14(256) + 5(16) + 11(1)$$
$$= 12288 + 3584 + 80 + 11$$
$$= 15963_{(10)}$$
$$= 15963$$

Note that to find these products, we have replaced E by 14 and B by 11. Recall that the first 16 decimal whole numbers 0, 1, 2, 3, 4, 5, 6, 7, 8, 9, 10, 11, 12, 13, 14, 15 are designated as 0, 1, 2, 3, 4, 5, 6, 7, 8, 9, A, B, C, D, E, F in the hexadecimal system.

Example

$$101111_{(2)} = \underline{1} \times 2^5 + \underline{0} \times 2^4 + \underline{1} \times 2^3 + \underline{1} \times 2^2 + \underline{1} \times 2^1 + \underline{1} \times 2^0$$
$$= 32 + 0 + 8 + 4 + 2 + 1$$
$$= 47_{(10)}$$
$$= 47$$

To generalize this procedure we shall have to use some algebraic notation. As you know, in algebra letters and other symbols (like x, y, z) are often used to stand for numbers; these symbols are called *variables* since their values are not determined until they have been definitely assigned. A detailed study of variables in general will be presented in Chap. 3. In many mathematical applications and particularly in certain types of computer programming, a whole set of variables will often have the same *common* name but be separately identified by subscripts. For example, we could choose the main variable name a and then write many other variables such as a_1, a_2, a_3, a_4 (which we read "a sub 1," "a sub 2," "a sub 3," "a sub 4")—all identified by the a but considered *different* variables because of the varying subscripts. In all our work, subscripts are always integers. A set of variables like $\{a_1, a_2, a_3, a_4, a_5, a_6\}$ is also called a *one-dimensional array*. The subscript itself could be a variable. We could let a_i stand for the general member of this one-dimensional array. We shall use subscripted variables from time to time, and in the chapters on computer programming we shall greatly expand the notion and applications of subscripted variables.

Let us now return to the problem of generalizing the procedure for determining the decimal value of a nondecimal number. Suppose we again let b stand for the base, let n stand for the number of digits in the integral part of the number, and let d_i stand for the digits in the number. We can then designate a number in base b as

$$d_{n-1}d_{n-2}d_{n-3} \cdot \cdot \cdot d_3d_2d_1d_0$$

The value of this number is given by

$$d_{n-1}b^{n-1} + d_{n-2}b^{n-2} + d_{n-3}b^{n-3} + \cdot \cdot \cdot + d_2b^2 + d_1b^1 + d_0b^0$$

Here the superscripts represent the usual algebraic exponents, that is, powers of the base b.

As you would expect, the decimal value of the fractional part of a number to a nondecimal base is determined by a process entirely analogous to the decimal case. For instance, suppose we have $0.2031_{(4)}$. We associate with each digit the positional value as shown below:

Number	2	0	3	1
Positional value	4^{-1}	4^{-2}	4^{-3}	4^{-4}

So, clearly,

$$0.2031_{(4)} = 2 \times 4^{-1} + 0 \times 4^{-2} + 3 \times 4^{-3} + 1 \times 4^{-4}$$
$$= \frac{2}{4} + 0 + \frac{3}{4^3} + \frac{1}{4^4}$$
$$= \frac{1}{2} + 0 + \frac{3}{64} + \frac{1}{256}$$
$$= \frac{128 + 12 + 1}{256}$$
$$= \frac{141}{256}$$
$$= 0.550781 \ldots$$

We note that the equivalent of $0.2031_{(4)}$ is *not* a terminating decimal. It is exactly $141/256$ but is approximately 0.5507 when truncated to four significant figures. Again we run into the problem of roundoff error when we attempt to store the base 4 number in a *decimal* register. Although 0.2031 will fit exactly into a base 4 four-digit register, the decimal value 0.5507 stored in a four-digit register has at least an error of 0.000081.

Example

$$0.10111_{(2)} = 1 \times 2^{-1} + 0 \times 2^{-2} + 1 \times 2^{-3} + 1 \times 2^{-4} + 1 \times 2^{-5}$$
$$= \frac{1}{2} + 0 + \frac{1}{8} + \frac{1}{16} + \frac{1}{32}$$
$$= 0.5 + 0.125 + 0.0625 + 0.03125$$
$$= 0.71875_{(10)}$$
$$= 0.71875$$

Example

$$B.3A_{(12)} = B \times 12^0 + 3 \times 12^{-1} + A \times 12^{-2}$$
$$= 11 \times 12^0 + 3 \times 12^{-1} + 10 \times 12^{-2}$$
$$= 11 + \frac{3}{12} + \frac{10}{144}$$
$$= 11 + 0.25 + 0.069(4)$$
$$= 11.319(4)$$
$$= 11.3194444 \ldots$$

Here 4 is an infinitely repeated digit. $B.3A_{(12)} \doteq 11.319$ (rounded to five significant figures).

EXERCISES

2.1 Determine the decimal equivalent (value) of the following:

(a) $4325_{(6)}$
(b) $11101111_{(2)}$
(c) $2AF3_{(16)}$
(d) $675_{(8)}$
(e) $120202_{(3)}$
(f) $1A31_{(11)}$
(g) $1234_{(5)}$
(h) $1234_{(6)}$
(i) $100143_{(5)}$

2.2 Determine the decimal value of the following. Express each exactly as a fraction, in the form of a decimal with repeating terminal digits, and also as a decimal truncated to five significant figures.

(a) $0.25_{(7)}$
(b) $0.010101_{(2)}$

(c) $0.30314_{(5)}$

(d) $0.0A35_{(12)}$

(e) $0.5667_{(8)}$

(f) $0.7DA_{(16)}$

(g) $0.10101_{(3)}$

(h) $0.23332_{(4)}$

(i) $0.0001101_{(2)}$

2.3 Determine the decimal value of the following. Express the fractional part as a repeating decimal.

(a) $102.12_{(3)}$

(b) $32.023_{(4)}$

(c) $34.221_{(5)}$

(d) $FF.FB_{(16)}$

(e) $173.465_{(8)}$

(f) $1101111.1101_{(2)}$

(g) $344.3_{(6)}$

(h) $50.2024_{(7)}$

(i) $ABC.91_{(13)}$

2.3 TRANSFORMING DECIMAL NUMBERS INTO NONDECIMAL NUMBERS

So far we have learned how to find the decimal value of a number in a non-decimal base. Now we present a method for changing decimal numbers into numbers in other bases. We will show the method in two parts — one for an integer decimal number and one for a purely fractional decimal number. Suppose we have a decimal integer N and desire to transform it into base b. Here is the general procedure. Let us divide N by b. Let q_1 be the quotient and r_1 the remainder. Now divide q_1 by b and let q_2 be the quotient and r_2 the remainder. Now divide q_2 by b and let q_3 be the quotient and r_3 the remainder. (Note the natural use we are making of subscripted variables to describe the procedure.) Now we continue in the same manner until the quotient is *zero*. Suppose that at that step the remainder is r_n. Then,

$$N = N_{(10)} = r_n r_{n-1} r_{n-2} \cdots r_3 r_2 r_{1\,(b)}$$

We note, of course, that the base b equivalent of the decimal integer N is composed of the remainders generated but in exactly the reverse of the order in which they were generated. The first remainder r_1 is the *last* digit of the number base b, and the last remainder r_n is the *first* digit. There are many ways to actually perform the mechanics of the steps just outlined. We have selected the schematic method shown below. In this model, the symbols represent the quantities we have just discussed.

b	N	q_1	q_2	q_3	q_4	$\cdots$	0
		r_1	r_2	r_3	r_4	$\cdots$	r_n

The arrow on the right side of the model is to emphasize that the remainders are to be taken in reverse order.

Example. Transform decimal 861 to base 5.

5	861	172	34	6	1	0
		1	2	4	1	1

Thus, $861_{(10)} = 11421_{(5)}$. Let us check this result:

$$11421_{(5)} = 1 \times 5^4 + 1 \times 5^3 + 4 \times 5^2 + 2 \times 5^1 + 1 \times 5^0$$
$$= 625 + 125 + 100 + 10 + 1$$
$$= 861$$

Example. Transform decimal 597 to base 11.

11	597	54	4	0
		3	A	4

Thus, $597_{(10)} = 4A3_{(11)}$. Note that the remainder r_2 is (decimal) 10 which is represented by A in the base 11 system. Now we again check:

$$4A3_{(11)} = 4 \times 11^2 + A \times 11^1 + 3 \times 11^0$$
$$= 4(121) + 10(11) + 3(1)$$
$$= 484 + 110 + 3$$
$$= 597$$

Example. Transform decimal 597 to base 2.

2	597	298	149	74	37	18	9	4	2	1	0
		1	0	1	0	1	0	1	0	0	1

Thus, $597 = 1001010101_{(2)}$.   Check:

$$1 \times 2^9 + 1 \times 2^6 + 1 \times 2^4 + 1 \times 2^2 + 1 \times 2^0 = 512 + 64 + 16 + 4 + 1$$
$$= 597$$

EXERCISES

2.4 Transform each of the decimal integers 0, 1, 2, 3, 4, 5, 6, 7, 8, 9, 10, 11, 12, 13, 14, and 15 into base 2.   Transform the decimal integers into base 4.   Repeat for base 8 and for base 16.

2.5 Use the method just presented to transform the following given decimal integers to the bases called for:

(a) 635 into base 2; into base 4; into base 5
(b) 87054 into base 8; into base 12; into base 16
(c) 9114 into base 3; into base 6; into base 9
(d) 450774 into base 4; into base 8; into base 16
(e) 67833 into base 3; into base 7; into base 11
(f) 1010101 into base 4; into base 5; into base 6

2.6 The number $837_{(9)}$ can be transformed into base 5 by first finding the decimal value of $837_{(9)}$ using the method of Sec. 2.1 and then transforming this *decimal* integer into base 5 using the method of this section.   Transform $837_{(9)}$ into base 5.

2.7 Following the plan of Exercise 2.6, transform the numbers below (with nondecimal bases) to the bases called for:

(a) $34421_{(5)}$ into base 6; into base 4; into base 8
(b) $11011101_{(2)}$ into base 4; into base 8; into base 16
(c) $4AF_{(16)}$ into base 3; into base 5; into base 8
(d) $4567A_{(11)}$ into base 15; into base 9; into base 8
(e) $3001223_{(4)}$ into base 8; into base 2; into base 6
(f) $44321_{(6)}$ into base 2; into base 4; into base 16

Let us now turn to fractional decimal numbers.   Again, let the number be n and the base be b.   *Multiply n by b.*   Let p_1 be the integral part of the product and f_1 be its fractional part.   Multiply f_1 by b.   Let p_2 be the integral part of the product and let f_2 be its fractional part.   Continue in the same manner as long as desired or until a repeating pattern emerges.   Then

$$n_{(10)} = .p_1 p_2 p_3 p_4 \cdot \cdot \cdot \quad (b)$$

Note that now the digits of the base b number are being generated by the procedure in exactly the same order as they occur in the number. Our schematic plan for this procedure is

b	n	f_1	f_2	f_3	f_4	f_5	$\cdots$
		p_1	p_2	p_3	p_4	p_5	

Example. Express decimal 0.4 in base 7.

7	0.4	0.8	0.6	0.2	0.4	0.8	0.6	$\cdots$
		2	5	4	1	2	5	

A repeating pattern occurs, as you can verify by proceeding. Thus, $0.4_{(10)} = 0.254125412541 \ldots_{(7)} = 0.(2541)_{(7)}$ using our repeating pattern notation for rational numbers. We note that 0.4 is terminating in base 10 but its equivalent base 7 is nonterminating.

Suppose we are using a computing machine that has been constructed to operate *internally* on base 7 numbers, and that it maintains three digits. In order to store the *external* exact decimal number 0.4, the best we could do is to *internally* store 0.254, the first three digits of the corresponding number base 7. Since the remainder of the base 7 number has been truncated, this introduces an error in merely *storing* of $0.00012541254 \ldots_{(7)} = 0.000(1254)_{(7)}$. We can compute this error in decimal form by evaluating $0.254_{(7)} = 2 \times 1/7 + 5 \times 1/49 + 4 \times 1/343 = 1/343(98 + 35 + 4) = 137/343$, so that the error is $0.4 - 137/343 = 2/5 - 137/343 = (686 - 685)/1715 = 1/1715$.

Example. Express the decimal 0.33 in base 2.

2	0.33	0.66	0.32	0.64	0.28	0.56	0.12	0.24	0.48	0.96	0.92	$\cdots$
		0	1	0	1	0	1	0	0	0	1	

Therefore $0.33 \doteq 0.0101010001_{(2)}$ (to 10 significant binary figures).

All current electronic digital computers are constructed to perform all their arithmetic operations on binary numbers. Thus, the remarks we have just made concerning a hypothetical computer which operated on base 7 numbers are particularly pertinent to actual computers. We here see that a computer which allows 10 binary digits cannot store the (external) decimal 0.33 exactly as a binary number. This means that the computer will, in general, be acting on approximate numbers.

If the decimal number has both nonzero integral and fractional parts, a combination of both parts of the method just discussed must be employed.

Example. Express decimal 37.51 in base 8.

8	37	4	0
	5	4	

Thus, $37_{(10)} = 45_{(8)}$.   And

8	0.51	0.08	0.64	0.12	0.96	0.68	$\cdots$
		4	0	5	0	7	

Thus, $0.51_{(10)} \doteq 0.40507_{(8)}$.   Finally, $37.51 \doteq 45.40507_{(8)}$.

EXERCISES

2.8 Express each of the following decimal numbers in the bases given. In each case, if the result is not terminating, carry to at least six figures or indicate the repeating pattern.

(a) 0.04375 into base 2; into base 4; into base 8
(b) 0.1 into base 12; into base 2; into base 5
(c) 0.443 into base 6; into base 5; into base 4
(d) 0.73 into base 2; into base 8; into base 16
(e) 0.005 into base 5; into base 8; into base 7
(f) 0.6245 into base 3; into base 12; into base 16

2.9 Express each of the following decimal numbers in the base given. In each case truncate the fractional part to five significant figures.

(a) 245.06 into base 2; into base 3; into base 4
(b) 4.44 into base 4; into base 8; into base 16
(c) 8764.55 into base 12; into base 13; into base 14
(d) 99.64 into base 6; into base 7; into base 8
(e) 16.1616 into base 4; into base 8; into base 16
(f) 983.17 into base 3; into base 9; into base 12

2.4 RELATED BASES

With each base b is associated a set of bases which is said to be related to base b. The related set of bases is composed of bases which are the various powers of b. For example, if $b = 2$, then the bases $4 = 2^2$, $8 = 2^3$, $16 = 2^4$, $32 = 2^5$, etc., are *related* to base 2. Similarly, if $b = 3$, then bases $9 = 3^2$, $27 = 3^3$, $81 = 3^4$, etc., are related. Because of their extraordinary importance, we shall be concerned here specifically only with the base 2 and its related bases. Consider the related base 4. The digits of the base 4 system are, of course, 0, 1, 2, and 3; these digits can be expressed as binary numbers of two digits, namely, 00, 01, 10, and 11, respectively. A number of base 2 can immediately be changed to base 4 by following this procedure: Mark off sets of two digits starting with the two preceding the decimal point and moving left. If the number has a fractional part, mark off sets of two digits starting from the decimal point and moving to the right. For example, $1\ 1\ \ 0\ 1\ \ 1\ 0\ \ 0\ 1\ \ 0\ \ 0_{(2)}$. Now replace each set of two binary digits by its decimal equivalent. In this case, we would have 3 1 2 1 0. This is the base 4 equivalent of the original binary number. We leave it to you to check this.

Example. Transform the binary number 1100110.111 into base 4. We mark off as prescribed: $1\ \ 1\ 0\ \ 0\ 1\ \ 1\ 0.\ \ 1\ 1\ \ 1\ 0$. Therefore, the result is $1212.32_{(4)}$. Note that the first pair is $0\ \ 1$.

Now consider the related base 8. The digits of the octal system are 0, 1, 2, 3, 4, 5, 6, and 7. Each one of them can be expressed as a binary number of three binary digits. These are, respectively, 000, 001, 010, 011, 100, 101, 110, and 111. A binary number can easily be transformed into base 8 by marking off (starting from the decimal point and moving left and right) sets of three binary digits. If each group is replaced by its decimal equivalent, the result will be the octal representation of the original binary number.

Example. $001\ \ \ 101\ \ \ 100\ \ \ 100_{(2)} = 1\ \ 5\ \ 4\ \ 4_{(8)}$.

Check: $1101100100_{(2)} = 2^9 + 2^8 + 2^6 + 2^5 + 2^2$
$$= 512 + 256 + 64 + 32 + 4$$
$$= 868_{(10)}$$
$$1544_{(8)} = 1 \times 8^3 + 5 \times 8^2 + 4 \times 8^1 + 4 \times 8^0$$
$$= 512 + 5(64) + 4(8) + 4$$
$$= 868_{(10)}$$

Base 16 is another important base related to base 2. As before, we now group in sets of four. Thus, $001101100100_{(2)} = 3\ \ 6\ \ 4_{(16)}$.

Check: $364_{(16)} = 3 \times 16^2 + 6 \times 16^1 + 4 \times 16^0$
$$= 3(256) + 6(16) + 4$$
$$= 868_{(10)}$$

Of course, base 16 has some nondecimal digits, so whenever the decimal equivalent of a set of four binary digits is more than 9, we replace it by the corresponding base 16 digit.

Example. $\underbrace{1111}\ \underbrace{1011}\ \underbrace{0111}\ \underbrace{0011}\ \underbrace{0001}\ .\ \underbrace{1101}\ \underbrace{1001}_{(2)}$ results in (decimal) 15 11 7 3 1. 13 9. Thus, FB731.D9 is the hexadecimal equivalent.

It should be clear from the discussion above that any number in a base related to 2 can very easily be expressed in base 2.   One simply replaces each of its digits by the corresponding binary equivalent.   Thus,

$$3\ 1\ 2\ 0\ 1_{(4)} = \underset{3}{\underbrace{11}}\ \underset{1}{\underbrace{01}}\ \underset{2}{\underbrace{10}}\ \underset{0}{\underbrace{00}}\ \underset{1}{\underbrace{01}} = 1101100001_{(2)}$$

and

$$A\ C\ 5\ E_{(16)} = \underset{A}{\underbrace{1010}}\ \underset{C}{\underbrace{1100}}\ \underset{5}{\underbrace{0101}}\ \underset{E}{\underbrace{1110}} = 1010110001011110_{(2)}$$

EXERCISES

2.10 Use the method of this section to transform the following binary numbers to base 4, to base 8, and to base 16:

(a) 11011011
(b) 11111110110111100
(c) 10100111101.1101
(d) 10101101001.10010011

2.11 Use the method of this section to transform the following numbers to base 2:

(a) $3223120_{(4)}$
(b) $FA8C_{(16)}$
(c) $0.77466_{(8)}$
(d) $1010337_{(8)}$
(e) $39.DE7_{(16)}$
(f) $3003312.2_{(4)}$

2.12 Transform $312103_{(4)}$ to base 16 by first transforming to base 2 and then transforming this result to base 16.

2.13 Recall that $16 = 4^2$ so that base 16 is related to base 4.   Transform $312103_{(4)}$ to base 16 by grouping its digits in sets of two, replacing each set (considered as a two-digit base 4 number) into decimal, and then replacing each decimal by its base 16 equivalent.

2.14 Since $9 = 3^2$, base 9 is related to base 3. Use the method described in Exercise 2.13 to transform $12102122_{(3)}$ into base 9.

2.5 ARITHMETIC OPERATIONS IN VARIOUS BASES

2.5.1 Addition

First, let us consider the basic operation of addition. The rule for addition of number base b is: Add the numbers of the rightmost column as usual using *decimal* addition. Transform this partial sum into its base b equivalent. The rightmost digit of this partial sum (in base b) is called the *sum digit*. Put down the sum digit in the sum as in ordinary addition and prepare for the *carry* operation. For example, if the partial sum is 67, then 7 is the sum digit and 6 is the carry digit; it would be carried and added to the next column to be added. However, if, for example, the sum is 112, then 2 is the sum digit, and the set of digits 11 are to be carried. These are carried over the next *two* columns, one in each column. After this, each column is added as described above.

Example. Add the following base 8 numbers:

```
7 3 4 2
6 7 0 7
5 5 5 4
3 0 6 2
```

Add the digits of the right column; the sum is 15 (decimal). This is $17_{(8)}$, and so 7 is the sum digit and 1 is the carry digit. Place the 7 and carry 1 to the next column. Add the digits of the next column. The sum is 16 (decimal). This is $20_{(8)}$. Place the sum digit 0 and carry 2. If you carry on in the same manner, the next partial sum is $17 = 21_{(8)}$ and the last partial sum is $23 = 27_{(8)}$. The final sum is $27107_{(8)}$.

Check:

$$
\begin{aligned}
7342_{(8)} &= 7(512) + 3(64) + 4(8) + 2 = & 3810 \\
6707_{(8)} &= 6(512) + 7(64) + 7 \quad\quad\;\; = & 3527 \\
5554_{(8)} &= 5(512) + 5(64) + 5(8) + 4 = & 2924 \\
3062_{(8)} &= 3(512) + 6(8) + 2 \quad\quad\;\; = & \underline{1586} \\
& & 11847_{(10)}
\end{aligned}
$$

and

8	11847	1480	185	23	2	0
		7	0	1	7	2	←

so that the result is $27107_{(8)}$, as above.

Example. Add the following base 5 numbers:

```
  13.214
 202.434
 123.433
 400.021
1300.212
```

In this problem, note that $12 = 22_{(5)}, 11 = 21_{(5)}, 10 = 20_{(5)}, 5 = 10_{(5)},$ and $8 = 13_{(5)}$.

Adding *two* binary numbers is especially simple. The following combinations are the only ones that can occur: $0 + 0 = 0, 0 + 1 = 1, 1 + 1 = 10,$ and $1 + 1 + 1 = 11$, all binary.

Example. Add the following two binary numbers:

```
 111011011
 101111011
1101010110
```

Finally, here is an interesting example: Add the following base 3 numbers:

```
      2
      1
     11
     11
   2122
   1022
  10012
  20122
  10002
  21021
  12122
1002201
```

The numbers that are written above the upper line are carry digits. The sum of the righthand column is $13 = 111_{(3)}$. Put down the 1 and carry 11 over the next two columns as shown. The second column adds to $12 = 110_{(3)}$. Put down 0 and carry 11 over the next two columns, as shown. The sum of the third column is $5 = 12_{(3)}$. Put down the 2 and carry 1 to the next column. The fourth column adds to $8 = 22_3$. Put down the 2 and carry 2. The fifth column adds to $9 = 100_{(3)}$. Finally, put down 100.

EXERCISE

    2.15 Add:

        (a) Base 6: 4033
                     2235
                     1052
                     2222

        (b) Base 9: 566778
                     345672
                     803366

        (c) Base 12: 455A2
                      98B39
                     345678
                     A00985

        (d) Base 16: 9 FDE
                     ABCD
                     3478
                     20F9
                     66A5

        (e) Base 2: 101110011
                     111011010
                     100011111
                     1011101110

        (f) Base 4: 32232
                     32221
                     3200022
                     3221102
                     321123

        (g) Base 5: 42310.23
                     23334.24
                     40003.01
                     23441.44

(h) Base 7:   5.66354
             4.62435
             3.33344
             1.00026
             3.55534

2.5.2 Subtraction

Next let us turn to the operation of subtraction. We have a choice of two methods, the first being the standard method we are all familiar with based on borrowing digits. The second method we shall present is to subtract by *adding* the $b - 1$ complement.

Example. Subtract using the method of borrowing (base 7):

$$
\begin{array}{r}
4025 \\
- \ 2346 \\
\hline
\end{array}
$$

Start from the right as usual. Since 6 is greater than 5, borrow 1 from 2 (leaving 1), add 7 to 5 and have 12. Now subtract 6 from $12 = 6$, the righthand digit of the remainder. Proceed to the next column. Again 4 is greater than 1, so borrow 1 from the 4, add 7 to the 0, borrow 1 from the 7 (leaving 6), and add 7 to $1 = 8$. Now subtract 4 from $8 = 4$. The rest of the subtraction proceeds without further borrowing. The remainder is $1346_{(7)}$. This example can be expressed as follows:

$$
\begin{array}{cccc}
 & ^6 & ^8 & \\
3 & \not{7} & \not{1} & 12 \\
\not{4} & \not{0} & \not{2} & \not{5} \\
\hline
2 & 3 & 4 & 6 \\
\hline
1 & 3 & 4 & 6
\end{array}
$$

in which the / indicates the borrowing maneuvers and the numbers above the upper line are part of the work involved.

Example. Subtract base 3:

$$
\begin{array}{r}
201210 \\
- \ 22122 \\
\hline
\end{array}
$$

Following the notation of the previous example, we have:

$$1\overset{2}{\cancel{3}}41\overset{3}{\cancel{0}}3$$
$$\cancel{2}\cancel{0}\cancel{1}\cancel{2}\cancel{1}\cancel{0}$$
$$-\ 22122$$
$$\overline{102011}$$

1781280

Check:

$$201210_{(3)} = 2(243) + 1(27) + 2(9) + 3 \qquad = 534$$
$$22122_{(3)} = 2(81) + 2(27) + 1(9) + 2(3) + 2 = 233$$

The difference is $301_{(10)}$. Now,

3	301	100	33	11	3	1	0
		1	1	0	2	0	1	←

Thus, $102011_{(3)}$.

Let us now turn to the second method for subtracting, the $b - 1$ complement method. It turns out that some version of this method is precisely the one used by digital computers to subtract, particularly binary numbers. If the base is b, first subtract each digit of the subtrahend from $b - 1$. Then *add* this result to the minuend. If a carry digit (defined below) appears, add this carry digit to the result; this process is often called an *end-around carry*. If no carry digit appears, the answer (the difference) is negative and is found by recomplementing the result. Examples will demonstrate the procedure.

Example. Decimal numbers, subtraction using the 9s complement method:

$$6132$$
$$-\ 4158$$

The 9s complement of 4158 is found by

$$9999$$
$$\underline{4158}$$
$$5841$$

Each digit is simply subtracted from $b - 1 = 9$. Now add this to 6132.

$$
\begin{array}{r}
6132 \\
5841 \\
\hline
\end{array}
$$

6132
5841
→1|1973

carry digit

(end-around   1
carry)

1974   result, which is easily checked

Example. Decimal numbers, using the 9s complement method:

    9247
− 15683

The 9s complement of subtrahend is clearly 84316.   Then

    09247
    84316
0|93563

There is no carry digit, so that the result is the 9s complement, that is, −6436. The result is negative.

Example. Base 5 numbers, by the 4s complement method:

    3240
− 1434

The 4s complement of 1434 is clearly 3010, the result of subtracting each digit in 1434 from 4.   Adding this to 3240, we have:

    3240
    3010
1|1300   addition is in base 5, of course
       1
    1301   result

Example. Base 16:

    639AF2
−   97C95

The 15s complement of 097C95 is F6836A.   Then

```
    639AF2
 +  F6836A
 1│5A1E5C
 ─────────
        1
   5A1E5D    result
```

Example. Base 7:

```
    02503
 −  61445
```

The 6s complement of 61445 is 05221.   Then

```
    02503
 +  05221
 0│11024
```

The result is, therefore, −55642.

The base 2 is of special interest.   First note that the 1s complement of a binary number is easily found.   Simply replace each 1 in the number by 0 and replace each 0 in the number by 1.   Let us check this.   Consider the number 11011.   Its 1s complement can be found by

```
    111111
 −  110011
 ─────────
    001100
```

and this is exactly the same as changing each 0 to 1 and each 1 to 0 in the original number.

Example. Base 2:

```
    1101100
 −  0110111
```

The 1s complement of 0110111 is 1001000.   Then,

```
  1101100
+ 1001000
1|0110100
        1
  110101     result
```

Example. Base 2:

```
  1101          001101
- 110011      + 001100
              0|011001
```

The result is -100110.

EXERCISES

2.16 Subtract using the direct (borrowing) method:

(a) Base 3:   21002211
            $-$ 2210121

(b) Base 2:   110101010
            $-$ 1010111

(c) Base 6:   430542
            $-$ 45323

(d) Base 8:   1170341
            $-$ 675473

(e) Base 14:   12 4 6 A 3
             $-$ BAC9D

(f) Base 16:   357 ADEA6
             $-$ 2FD7 7 68 E

2.17 Subtract using the $b-1$ complement method:

(a) Base 2:   1101101
            $-$ 1010010

(b) Base 3:   21101
            $-$ 222102

(c) Base 7:   4023561
            $-$ 2606635

(d) Base 11: 345678
            − 4963A1

(e) Base 16: 1234567
            − ABCDEF

(f) Base 8: 132435672
           − 24566777

(g) Base 4: 3030303
           − 3301223

(h) Base 10: 98076543
            − 123677898

2.5.3 Multiplication

Now for the operation of *multiplication*. Perhaps a detailed, worked-out example would be instructive. Consider this problem: Multiply, using base 6, 325×43.

```
   325
 ×  43
  1423
  2152
 23343
```

Analysis: First place the numbers under each other as in usual multiplication in decimal system. Multiply 3 by 5 as decimals. The result is 15 which we place in base 6 as 23. Put down the 3 and carry the 2. Now $2 \times 3 = 6, 6 + 2 = 8$ (decimal). Place 8 in base 6 as $12_{(6)}$. Put down the 2 and carry the 1. Now $3 \times 3 = 9$, $9 + 1 = 10$, and $10_{(10)} = 14_{(6)}$. As shown, the first intermediate product is 1423. Shift one to the left as is usual in this type of multiplication. Multiply 4 by 5. Result is $20 = 32_{(6)}$. Put down the 2 and carry the 3 as expected. Now $4 \times 2 = 8$, $8 + 3 = 11$, and $11 = 15_{(6)}$. Put down 5 and carry 1. Then $4 \times 3 = 12$, $12 + 1 = 13 = 21_{(6)}$. This completes the second intermediate product as $2152_{(6)}$. Now add base 6. The result, the product, is $23343_{(6)}$.

   We check this result. First, we leave it to you to show that $325_{(6)} = 125_{(10)}$ and that $43_{(6)} = 27_{(10)}$. Then we multiply these two base 10 equivalents for a product of 3375. Now

6	3375	562	93	15	2	0
		3	4	3	3	2	←

Thus, $3375 = 23343_{(6)}$ and the check is complete.

Example. Multiply the base 3 numbers 21022 and 2112.

$$
\begin{array}{r}
21022 \\
2112 \\
\hline
112121 \\
21022 \\
21022 \\
112121 \\
\hline
200101011
\end{array}
$$

Check the details of this multiplication; recall that $3 = 10_{(3)}$, $4 = 11_{(3)}$, $5 = 12_{(3)}$, $6 = 20_{(3)}$, and $7 = 21_{(3)}$. Verify the result by showing that $21022_{(3)} = 197_{(10)}$, $2112 = 68_{(10)}$, the decimal product is 13396, and $13396_{(10)} = 200101011_{(3)}$.

Example. Multiply, base 16:

$$
\begin{array}{r}
3A7 \\
\times\ B9 \\
\hline
2\,0DF \\
28\,2D \\
\hline
2A3AF
\end{array}
$$

Again, you should check the multiplication itself, and if you care to, verify the result by following the steps as outlined above.

Example. Multiply, base 2:

$$
\begin{array}{r}
11011 \\
\times\ 11101 \\
\hline
11011 \\
11011 \\
11011 \\
11011 \\
\hline
1100001111
\end{array}
$$

The partial products in this example are, of course, very easy to calculate. Some intricacies occur during the addition of these partial products using binary addition. When the (decimal) sum of a column is 4, the binary equivalent is 100. The right zero is placed down, and the left 10 is carried over the next *two* columns to the left. Check this example carefully.

EXERCISE

2.18 Multiply using the method described in the text:

(a) Base 4: 13013
            × 231

(b) Base 9: 483
            × 78

(c) Base 11: 4A3
             × 9A

(d) Base 2: 1011101
            ×   1011

(e) Base 8: 4775
            × 1247

(f) Base 16: BEF
             × AC

(g) Base 5: 4432
            × 302

(h) Base 13: 5BA8
             × 4C

2.5.4 Division

The operation of division is straightforward like ordinary division but carrying it out becomes a rather tedious experience. It necessarily makes use of the previously discussed operations of subtraction and multiplication. Let us try an example and discuss it in some detail.

Example. Divide, base 5, 123221 by 32.

$$32 \overline{)\ 123221}$$

Following the usual division procedure, we attempt the trial division of 32 into 123. First, try 3. But $32 \times 3 = 201_{(5)}$, and so 3 is too large. Next, try 2. $32 \times 2 = 114_{(5)}$. This is satisfactory, so place 114 under 123 and subtract (base 5):

```
          2
32 ) 123221
     114
      42
```

Carry down the next digit as usual. Clearly 32 goes into 42 once. We now have

```
         21
32 ) 123221
     114
      42
      32
     102
```

Clearly, 32 goes into 102 only once, so we now have

```
        211
32 ) 123221
     114
      42
      32
     102
      32
     201
```

Luckily, 32 goes into 201 three times evenly. We have finally

```
       2113
32 ) 123221
     114
      42
      32
     102
      32
     201
     201
       0
```

Example. Divide 1F8) 734AC (base 16).

```
              3A8
       1F8 ) 734AC
             5E8
             14CA
             13B0
             11A C
              F C0
              1 E C      remainder
```

This example is, of course, the result of lots of trial and error in the trial divisors. We have chosen to stop the division as shown and indicate the remainder. One could have supplied a decimal point and continued by inserting zeros in the dividend. We check the result of this division problem.

```
        1F8
      × 3A8
        FC0
     13 B0
      5E8
     732 C0
   +   1 E C      remainder
     734 AC      check
```

Although the method of division we have just shown is loads of fun to do (as well as a test of patience), most computers (either electric or electronic) do not perform the division operation in this way. Instead they use a method of repeated subtraction as demonstrated in the following example.

Example. Divide 8041 by 17 (base 10) using the method of repeated subtraction.

First we set up a four-digit register in which to store the quotient. In addition to the ordinary process of subtraction, we shall require that the computer (or you if you are doing the division by hand) be able to count how many times the divisor 17 is subtracted during the process. This means that each position in the four-digit register we have set up is actually a counter. Here is how we proceed. To determine the first digit in the quotient, take the first digit of the dividend and subtract 17 (the divisor):

```
    8
 − 17
   −9
```

Since the remainder −9 is negative, do not advance the first counter by 1 but move one place to the right in the dividend and consider the first two digits 80. At the same time advance to the second counter. Again subtract 17, over and over again as follows:

$$
\begin{array}{r}
80 \\
-17 \\
\hline
63
\end{array}
$$

Since the difference is positive, advance the counter 2 by 1. The register now looks like this: $\boxed{0\mid1\mid0\mid0}$. Proceed (using the second counter):

$$
\begin{array}{r}
63 \\
-17 \\
\hline
46 \\
-17 \\
\hline
29 \\
-17 \\
\hline
12 \\
-17 \\
\hline
-5
\end{array}
$$

(advance counter 2 by 1)

(advance counter 2 by 1)

(advance counter 2 by 1)

Since this last remainder is negative, do not advance the counter. The register now looks like this: $\boxed{0\mid4\mid0\mid0}$.

Now move back to the *previous* remainder 12, bring down the next digit of the dividend, and advance to the third counter. Proceed as before:

$$
\begin{array}{r}
124 \\
-17 \\
\hline
107 \\
-17 \\
\hline
90 \\
-17 \\
\hline
73 \\
-17 \\
\hline
56 \\
-17 \\
\hline
39 \\
-17 \\
\hline
22 \\
-17 \\
\hline
5 \\
-17 \\
\hline
-12
\end{array}
$$

(advance counter 3 by 1)

(advance counter 3 by 1)

(advance counter 3 by 1)

(advance counter 3 by 1)

(advance counter 3 by 1)

(advance counter 3 by 1)

(advance counter 3 by 1)

(do not advance counter 3)

Recover the previous remainder 5, carry down the next digit of the dividend, move to counter 4, and proceed, starting with the number 51. If one carries this procedure through the necessary steps to find the final value of counter 4, the quotient register will then contain $\boxed{0\;|\;4\;|\;7\;|\;3}$; the quotient is 473.

Before we turn to some discussion of this method, let us present another example which is closer to the way computers actually work. In this example, the subtractions that are required over and over again will be done using the method of adding the $b-1$ complement.

Example. Divide, base 4, 123103 by 321. The 3s complement of the divisor 0321 is 3012. We shall set up a six-digit quotient register of counters. Clearly, the first three digits of the quotient are 0 0 0. Proceeding from there, we have:

```
           0321 )  123103
                  + 3012
(end-around     1│ 0303
   carry)            ⌐→ 1
                 ─────────
                   0310      (advance counter 4 by 1)
                  + 3012
                 ─────────
                   3322
```

Recall that since the last addition results in no carry digit, this means that the last difference is actually -0011. We do not advance counter 4 but instead recover the previous difference 310, bring down the next digit of the dividend, proceed to counter 5, and restart the process on the number 3100. The remainder of the process goes like this:

```
   3100
 + 3012
 1│2112
    ─ 1
 ───────
   2113      (advance counter 5 by 1)
 + 3012
 1│1131
    ─ 1
 ───────
   1132      (advance counter 5 by 1)
 + 3012
 1│0210
    ─ 1
 ───────
   0211      (advance counter 5 by 1)
 + 3012
 0│3223      (do not advance counter 5)
```

So far the quotient register looks like this: | 0 | 0 | 0 | 1 | 3 | 0 |. Recover the previous difference 211, carry down the next digit of the dividend, advance to the next counter, and proceed.

$$
\begin{array}{r}
2113 \\
+\,3012 \\
\hline
1\,|\,1131 \\
1 \\
\hline
1132 \\
+\,3012 \\
\hline
1\,|\,0210 \\
1 \\
\hline
0211
\end{array}
$$

     (advance counter 6 by 1)

     (advance counter 6 by 1)

Clearly, repeating will lead to a negative remainder. We conclude now with the quotient register looking like this: | 0 | 0 | 0 | 1 | 3 | 2 | and with remainder 211.

Considering all the work involved in the two previous examples, you can see why one does not employ this method when computing by hand. But an electronic computer can perform at a dizzying pace. It performs *all* arithmetic operations using only one basic operation — addition. We have seen that we can subtract by adding, for example, the $b-1$ complement. Clearly, multiplication can be achieved by repeated addition. And now we see how division can be done using only the arithmetic operation of addition. The computer does not mind doing all the required additions, the setting up and advancing of counters in registers, the shifting left or right as required by the operations of multiplication and division. Circuits have been devised to do all these things and myriads of other required tasks. As far as arithmetic operations are concerned, it is much simpler to make the computer able to do only one operation but at a fantastically rapid pace. The circuits required to perform addition are very simple to design. To have separate circuits designed for each operation would terrifically complicate the computer.

EXERCISES

2.19 Divide using the direct method. Carry to at least one fractional digit if the division is not even. State the remainder in each case.

    (a) Base 5: $23\,)\,\overline{14302}$
    (b) Base 2: $1011\,)\,\overline{11011101.1}$
    (c) Base 7: $43\,)\,\overline{61405}$
    (d) Base 16: $FA\,)\,\overline{19E47}$

(e) Base 4: $322\overline{)3330231}$

(f) Base 8: $67\overline{)171717}$

2.20 Divide using the method of subtraction as described in the text. Actually subtract, do not use the method of adding the $b-1$ complement.

(a) Base 2: $101\overline{)1110111}$
(b) Base 8: $347\overline{)42562}$
(c) Base 4: $322\overline{)332021}$
(d) Base 16: $9F\overline{)C37DE}$

2.21 Divide using the method of subtraction. Do the subtracting by using the method of addition of the $b-1$ complement.

(a) Base 2: $101\overline{)110111}$
(b) Base 7: $346\overline{)256332}$
(c) Base 12: $98A\overline{)B345B}$
(d) Base 10: $756\overline{)568934}$

2.6 BINARY CODED DECIMAL NUMBERS

Another method sometimes used to translate external decimal numbers into a binary form that the computer can easily manage is called *binary coding*. Although we shall present this method for hand calculations, you should realize that the coding and the subsequent operations of the computer on these coded numbers are handled completely by the computer itself internally.

First, recall (see Exercise 2.4) that each of the decimal numbers 0 through 15, inclusive, can be expressed in binary form using four binary digits:

0	0000	4	0100	8	1000	12	1100
1	0001	5	0101	9	1001	13	1101
2	0010	6	0110	10	1010	14	1110
3	0011	7	0111	11	1011	15	1111

Thus, 16 four-digit binary numbers can be constructed. Now suppose that one associates one of these 16 numbers with each decimal *digit*, but *not* necessarily the binary equivalent of the decimal digit. There are 10 decimal digits, and so evidently this association can be done in quite a number of ways. Here is one way which we shall call *code 1* (each set of associations is called a *coding*):

0	0011	5	0100
1	1001	6	0010
2	0001	7	1000
3	1111	8	1010
4	1101	9	1110

You note, of course, that this particular coding does not associate any decimal digit with its corresponding binary equivalent. Any decimal number can now be coded using this code by following this rule: Replace each digit of the decimal number by the corresponding four-digit binary number according to the code.

Example. 8 6 2 (decimal number) is coded thus:

    1010   0010   0001

The 12-digit coded result 101000100001 is, naturally, *not* the ordinary binary equivalent of the decimal number 862. [$862_{(10)} = 1101011110_{(2)}$]. Thus, as you can see, once the code has been selected, it is very easy to code any decimal number.

The problem of consequence arises once the computer has coded a set of decimal numbers and then must perform the operation of addition upon them, retranslate the sum to ordinary decimal, and output the result. To see this difficulty more clearly, suppose we code the decimal numbers 2 and 4 using code 1 and add these numbers using binary addition. We have

2	0001
4	1101
	1110

The result is 1110. According to our code this corresponds to decimal 9. But clearly 9 is not the sum of 2 and 4. In a rather arbitrary coding like this one we have just chosen, there is no simple way to figure out just how we would ever make the result 0010, which is the code for decimal 6, the proper result.

A commonly used code which we shall call *code 2* associates each decimal digit with its four-digit binary equivalent. That is,

0	0000	5	0101
1	0001	6	0110
2	0010	7	0111
3	0011	8	1000
4	0100	9	1001

We shall now proceed to investigate this code. *Other* codes are in actual use in some computers.

One would think that all the problems that arose in code 1 would now evaporate. However, consider

```
 6  0110
 5  0101
11  1011
```

The correct result is decimal 11, a number of *two* digits, while the binary addition of the coded decimals is of four digits, cannot represent a two-digit number, and besides is not one of the allowable four-digit numbers of code 2. But notice this. Suppose that to this result (1011) we add 0110, which is 6 in code 2.

Thus,

```
6        0110
5        0101
         1011
      +  0110      (binary equivalent of decimal 6)
      00010001
         1   1
```

We have inserted three zeros to the left of the first 1 to complete the four-digit requirement of the code. Then, as if by magic, after translating back to decimal digits using the code, we have decimal 11, the correct result. This can be expressed in the following rule: If the sum of the coded digits is more than the binary equivalent of 9, add 0110 to the sum, then decode as shown above.

Let us now try adding two decimal numbers of several digits.

Example. Add

```
8  6  9       1000  0110  1001
          →
4  8  7       0100  1000  0111
```

We have shown the code 2 coding of the given decimal numbers slightly spread apart for clarity. The procedure is to now add (using binary addition) but to be careful to note whenever the sum of a set of two numbers is greater than the binary equivalent of 9. Otherwise, we add as usual carrying carry digits as we ordinarily do. This is what occurs:

8 6 9		1000	0110	1001
4 8 7		0100	1000	0111
13 5 6		1100#	1111#	0000#
		0110	0110	0110
	0001	0011	0101	0110
	↓	↓	↓	↓
	1	3	5	6

The symbols (#) after the sums of the third row indicate that binary 0110 is to be added to them since the sum of the corresponding set of two binary numbers is more than the binary equivalent of 9. The second addition operation is done as usual with carry digits if required. The first set 0001 is completed as before. In any case, the set of four four-digit binary numbers corresponds, by recoding back to decimal, to the proper answer.

Example

4 3 9 2		0100	0011	1001	0010
8 1 7 6		1000	0001	0111	0110
12 5 6 8		1100#	0101	0000#	1000
		0110		0110	
	0001	0010	0101	0110	1000
	1	2	5	6	8

We shall not proceed any deeper into the further intricacies of the properties of binary coding of decimal numbers. Perhaps you can figure out, for this particular coding, why the addition of 0110 to the proper sums makes the process work.

EXERCISES

2.22 Construct a binary code different from the two used in this section and use it to code the following decimal numbers:

(a) 8142
(b) 93047
(c) 63415
(d) 3999

2.23 Use code 2 of this section to code the following decimal numbers, then add them using the rules given. Follow the presentation plan used in the text, that is,

	Decimal	*Binary coded*		
	8 2	1000	0010	
	4 7	0100	0111	etc.
	12 9			

(a) 7
     5̲

(b) 75
     2̲3̲

(c) 75
     4̲8̲

(d) 325
     2̲9̲9̲

(e) 1473
     9̲2̲4̲3̲

(f)   2837
     4̲8̲7̲9̲6̲

(g) 8007
     3̲9̲9̲8̲

(h) 437 ←
     245 ←
     9̲3̲8̲ ←

3

Algebra: Linear Functions

3.1 PRELIMINARY REMARKS

In algebra we construct mathematical expressions and statements using two kinds of symbols representing numbers. We use *constants* (or explicit numbers) like 7, −34, and 5.62 — these symbols are fixed in value; we use *variables* (or literal numbers) like a, b, x, y, α, and θ — these are nonnumeric symbols, usually letters of the English or Greek alphabet. The variables stand for numbers but may take on different values in varying contexts. All these symbols, constant or variable, either are numbers or represent numbers. They can be combined following simple rules by using the four operations of addition, subtraction, multiplication, and division into mathematical phrases called *expressions*. Standard symbols are used to represent these operations. For addition, we use +. Examples of expressions using + are $a + 3$, $7 + 9.3$, and $x + y$. For subtraction we use −. Examples are $a - 3$, $7 - 9.3$, and $479 - y$. The symbols representing multiplication are various. Sometimes we use × or ·. Examples of expressions using these symbols for multiplication are $a \times b$, $473 \times y$, and $c \cdot d$. Often in algebra we simply omit either symbol and indicate multiplication by placing the numbers or symbols directly next to each other. For example, $4y$ means $4 \times y$, and abc means $a \times b \times c$. Of course, we cannot do this if the two symbols are both constants. Would you interpret 738 as meaning 7×38? In the product of a constant and a variable, we always write the constant first. Thus, it is $4y$ and not $y4$. This is an example of algebraic etiquette. As you recall, if a variable or constant is multiplied by itself, we use the symbol representing power or exponentiation. Thus, we write, if we want to, $a \times a$ as a^2, $a \times a \times a$ as a^3, and so forth. As we shall see in the next chapter, the symbolic programming language called Fortran uses the asterisk * as the symbol for the operation of multiplication. There we shall write A * B instead of $a \times b$ (or ab). Note also the use of nonitalic capital letters in this example. For the operation of division the symbols / or —— are the most common. The symbol ÷ used sometimes in arithmetic is rarely used. Some examples of expressions using division symbols are a/b (a divided by b), $\frac{x}{7}$, and 4.5/3.7. The slash symbol / is the one most used in computer languages. In Fortran we write such expressions as X/Y and A/2.5.

Of course, many expressions contain several of these operation symbols. Thus, $x + 3y$, $a - c/d$, and $a + b - 4x$ are properly formed expressions. An expression can be evaluated if the values of the variables it contains are specified. Suppose x has the value 7 and the variable y has the value 8. Then the expression $x + 3y$ has the value 31 (that is, $7 + 24$). If the variable x has the value 8.7 and y has the value -3.2, then the same expression $x + 3y$ has the value -0.9. We find this value by substituting the given values for x and y into the expression and calculating the value by evaluating $8.7 + (3)(-3.2)$.

When an expression contains more than one operation, a definite order in which the operations are to be performed has been agreed upon. This order is called the *hierarchy of operations* and is as follows:

1. Exponentiation
2. Multiplication and division
3. Addition and subtraction

Let us now consider some expressions that do not contain parentheses and determine the order of performance of the operations they contain.

Example. Suppose a is 4, b is -3, and c is -7. Then the expression $ab^2 + c$ is evaluated like this: Since it contains exponentiation, that operation is done first; b^2 is evaluated as 9. It contains the product of a by b^2, so the multiplication is done next: the product of a and b^2 is 4×9 and has value 36. Since there are no more multiplications or divisions, we start executing any additions or subtractions involved; to the product 36 we now add the value of c. We finally evaluate the numerical expression $36 + (-7)$ and arrive at the final result of 29, the value of the expression.

Example. Evaluate the expression $ab + c$ if a is 12, b is 3, and c is 5.

Solution: Multiply a by b, then to this product add c.   The result is $36 + 5$, so the value of the expression is 41.

Example. Evaluate the expression $a/b - c/d$ if a is 6, b is 4, c is 5, and d is 4.

Solution: Divide a by b, divide c by d, then subtract these two quotients. Since a/b is 1.5 and c/d is 1.25, the value of the expression is 0.25.

The symbol $=$ (read "is equal to") is used to equate two expressions. The resulting algebraic statement is called an *equation*.   Thus, the equation $x + y = 7$ is read "x plus y equals 7" or "the sum of x and y is 7."   Note the difference between an expression and an equation.   An algebraic expression is a mathematical *phrase*; an algebraic equation is a mathematical *sentence*.

An equation is said to be *satisfied* for a certain set of values of the variables it contains if the two expressions equated become identical when the values of the variables are substituted into them.   For example, the equation $x + y = 7$ is satisfied if $x = 5$ and $y = 2$.   It is not satisfied for $x = 3$ and $y = 8$. Evidently many sets of values for x and y satisfy the equation $x + y = 7$.   As you recall, many equations are satisfied by a unique value of the variable contained.   Thus, $x + 3 = 8$ is satisfied only when $x = 5$.   In such an equation, finding the unique value that satisfies it is called *solving* the equation.

EXERCISES

3.1 Evaluate each of the following expressions for the values of the variables given:

(a) $ax - 7$, $a = 3$, $x = 7$

(b) $a/b + cd$, $a = 3$, $b = 2$, $c = 4$, $d = 1.25$

(c) $2a^3 - \dfrac{c}{d}$, $a = -2$, $c = 12$, $d = -6$

(d) $\dfrac{c + d}{a - b}$, $a = 12$, $b = 0$, $c = 3$, $d = 9$

(e) $4x^2 - 8x + 5$, $x = \frac{1}{4}$

(f) $a - \dfrac{b}{c} + d$, $a = 2$, $b = 6$, $c = 3$, $d = 5$

(g) $x^2 + y^2 - z^2$, $x = 2$, $y = 3$, $z = 4$

(h) $\dfrac{a^3}{5} + xy$, $a = 10$, $x = -6$, $y = 30$

3.2 Solve each of the following equations:

(a) $3x + 7 = 7$

(b) $2 - x = 8$

(c) $3a = 5 + 7$

(d) $\dfrac{x-6}{5} = 3$

(e) $2 + 3x = 5 - x$

(f) $\dfrac{x+7}{3} = \dfrac{x-5}{6}$

3.3 Translate each of the following word statements into algebraic equations.

(a) The sum of p and q is -8.
(b) The product of a and b is 12.
(c) The product of 4 and x added to the product of 3 and y is 15.
(d) The sum of a and b is equal to the product of c and d.
(e) The square of a is 37.
(f) The area A of a rectangle is the product of its length L and its width W.
(g) The distance d is the product of the velocity v and the time t.

3.4 (a) Find at least five sets of values for x and y each of which satisfies the equation $x + 3y = 8$.
(b) Solve the equation $x^2 = 16$.
(c) Show that $x = 2$ does not satisfy the equation $x^3 + 7x^2 - 8x - 67 = 0$.
(d) If $x = 0$, $y = 7$, evaluate the expression xy, the expression $x + y$, the expression y/x, the expression $3x^2 + 7xy + 8y^2$.

3.5 Solve the following linear equations for x if all the constants and the variable x are in the given base. All operations required to solve for x are thus to be *performed in that base*.

(a) Base 5: $3x - 4 = 131$
(b) Base 8: $7x + 11 = 333$
(c) Base 4: $10x + 201 = 2010 - 13x$
(d) Base 12: $63 = 1A3 - 14x$

Another set of symbols often used in algebraic expressions is the pair of parentheses (and) called, respectively, the opening parenthesis and closing parenthesis. These are symbols of *grouping* and indicate that the quantities they enclose are to be considered as a unit. They always occur in pairs; for each opening parenthesis there must be an accompanying closing parenthesis.

Their use in an expression can alter the order in which operations are done. Here are some examples of expressions containing parentheses.

$$a + (b + c)/d \quad (a + c)/d \quad (a + b)/(c + d)$$
$$a + (b/c) + d \quad a + b/(c + d) \quad (a + d(b + c))/e$$

The last example contains what are called *nested parentheses*, one set of parentheses within another set. In such an expression, we follow this rule: To evaluate an expression containing parentheses, proceed to the innermost parentheses and evaluate the expression they contain according to the usual hierarchy rules. Then move to the next outer pair of parentheses, evaluate, and so on. Consider the expression $(a + d(b + c))/e$. Since the innermost parentheses enclose $b + c$, we calculate this sum first. We then proceed to the outer pair, multiply $b + c$ by d, and add the result to a. Finally we divide by e. For example, if $a = 3, b = 6, c = 8, d = 4$, and $e = 2, b + c = 14, d(b + c) = 4(14) = 56$, $a + d(b + c) = 3 + 56 = 59$, and $(a + d(b + c))/e = {}^{59}\!/_2 = 29.5$. In algebra other symbols of grouping such as brackets [ ] and braces { } are sometimes used. In most computing applications the parentheses are the only symbols of grouping that are permitted, no matter how nested they have to become. In addition to possibly redirecting the order of operations, parentheses can be used to provide clarity even when they are not actually required. It is perfectly permissible to write $(ab) + (c/d)$ even though $ab + c/d$ means exactly the same thing. However, some expressions cannot be written without using parentheses. For example, to divide r by the quantity $a + b$, we must write $r/(a + b)$; we cannot write $r/a + b$.

EXERCISE

3.6 If $a = 2, b = 3, c = 8$, and $d = 2$, evaluate each of the following expressions:

(a) $a + b/c$
(b) $(a + b)/c$
(c) $(a + b)/c + d$
(d) $(a + b)/(c + d)$
(e) $a + (b + c)d$
(f) $a - b/(c + a)$
(g) $(a - b)/c + d$
(h) $(a - b)/(c + d)$
(i) $(3a - 4b)/(bc)$
(j) $(3a - 4b)/bc$
(k) $12a/bc$
(l) $12/1/b$

(m) $(3a - b)^2/d$

(n) $3a - b^2/d$

(o) $(a + b)^2/(a - b)^2$

(p) $4c^2/d^3 + c/d + 7/(b + 2d)$

3.2 FUNCTIONS

When an equation involves one or more variables, it can be solved using standard algebraic operations for one of them. Thus, the equation

$$3x + 7y = 8 + w \tag{3.1}$$

contains the variables x, y, and w. Suppose we solve for y. This means that we want to, without in any way altering the mathematical content or meaning of the equation, write the equation in the form $y = \ldots$. Clearly, $7y = 8 + w - 3x$, and $y = (8 + w - 3x)/7$. (Note the use of parentheses; the result is *not* $y = 8 + w - 3x/7$.) The value that y will have is determined uniquely if specific values are given to w and x. We say that the variable y is a single-valued *function* of the variables x and w. Since the value of y is dependent upon the values assigned to x and w, the variable y is called the *dependent* variable. The two variables x and w are *independent* variables since, presumably, any values may be assigned to them.

A short glance back to Eq. (3.1) is sufficient to make one realize that any of the three variables it contains can be regarded as a function of the other two. We can solve for w explicitly. Thus, $w = 3x + 7y - 8$. Here w is the dependent variable and x and y are independent. Again, w is a single-valued function of x and y since it is clear that there is one and only one value of w for each selection of the pair of values for x and y.

It is conventional to represent functions by a certain notation called, appropriately enough, *functional notation*. We write, for example, $y = f(x)$ (read "y is f of x" or "y is a function of x, the function is named f") to express the fact that the value of the dependent variable y depends on the independent variable x. Note the use of parentheses to enclose the x, often called the *argument* of the function. The equation $y = f(x)$ does not give explicitly the procedure one is to follow to determine y if a value of x is prescribed. When one actually substitutes a specific value of x into the expression $f(x)$, one denotes this by replacing the x within the parentheses by the specific value of x. For example, if $x = a$, we write $f(a)$; if $x = -2$, we write $f(-2)$. Suppose $f(x) = x^2 - 7x + 3$. Then $f(3) = 3^2 - 7(3) + 3 = 9 - 21 + 3 = -9$, and $f(-b) = (-b)^2 - 7(-b) + 3 = b^2 + 7b + 3$. Similarly, $f(2/x) = (2/x)^2 - 7(2/x) + 3 = 4/x^2 - 14/x$

+ 3.   The letter f in $f(x)$ is not sacrosanct.   Other letters such as g, h, F, G, or H can be used.

A similar notation is used for functions of several variables.   Suppose $z = g(x,y,p) = x + 3y - 7p$.   Then $g(2,3,7) = 2 + 3(3) - 7(7) = 2 + 9 - 49 = -38$.   Clearly the notation $g(2,3,7)$ means that 2 is to be substituted for x, 3 for y, and 7 for p wherever these variables occur in the function expression.

Example. Let $h(a,b,\theta) = \dfrac{a^2 + b^2}{2a\theta}$.   Then $h(1,3,p) = \dfrac{1^2 + 3^2}{2(1)p} = \dfrac{10}{2p} = \dfrac{5}{p}$.

EXERCISES

3.7   Given $3x + 4y - 5q = 8$, find x as a function of y and q; find y as a function of x and q; find q as a function of x and y.

3.8   Given $x^2 - y = 6$, find y as a function of x; show that x is *not* a single-valued function of y.

3.9   If $f(x) = 4 - x^2$, find $f(1)$; $f(\sqrt{s})$; $f(2)$; $f(t^3)$.

3.10   If $f(x) = x^2 + 5$, find $f(x + h)$; $f(x + h) - f(x)$.

3.11   If $g(p) = \dfrac{p - 4}{5 + p}$, find $g(5)$; $g\left(\dfrac{2}{p}\right)$; $g(s - 5)$; $g(4)$; $\dfrac{4}{g(r)}$.

3.12   If $h(x,y) = x - 4y + 1/x$,   find $h(1,2)$; $h(\frac{1}{2},-2)$; $h(1/a,1/b)$; $h(s + 1,s + 3)$; $h(3,0)$.

3.13   If $G(r,s,t) = r^2 + s^2 + t^2$,   find $G(0,0,2)$; $G(2,3,-4)$; $G(1,0,\sqrt{s})$; $G(\frac{1}{2},\frac{1}{2},\frac{1}{4})$; $G(-1,-2,-3)$; $G(\sqrt{3},\sqrt{5},\sqrt{6})$.

This concludes our introduction to functions.   In the remainder of this chapter we shall restrict ourselves to *linear* functions.   A function (of any number of variables) is called linear if each of the terms it contains is of first degree or is a constant.   Thus, $8x - 7y + 9z$ is a linear function of x, y, and z, and $56 + a + 3b$ is a linear function of a and b.   The simplest linear function, other than a single constant, is one containing only one variable, for example, $3x + 7$, $-41 + 17.3r$, and $8\phi - 45$.   It is nothing short of amazing how important these rather simple kinds of functions are in all sorts of significant applications.   We shall study some of these applications of linear functions in this chapter.

3.3 COORDINATE SYSTEMS. ANALYTIC GEOMETRY OF THE STRAIGHT LINE

In order to visualize geometrically the linear function of one variable, let us choose x as the independent variable, y as the dependent variable, so that $y = f(x)$ where $f(x)$ is a linear function of x. We now set up a (one-dimensional) coordinate system on which we shall represent the values of the variable x. First, draw a horizontal line and label it with the symbol x as shown in Fig. 3.1. Now mark a point on the line and associate the real number zero with this point. This point will be called the *origin*. If you then mark other points using a uniform scale as shown in Fig. 3.1 and presume that the line extends indefinitely to the left and right, you can see that all the values that the real variable x can take on are associated with unique points on the line. And, alternately, each point on the line has associated with it a unique real number. The line has now become a one-dimensional coordinate system called the x axis. Now for the dependent variable y. We copy the same coordinate system with exactly the same set of real numbers on it, but we label the variable y instead of x. Clearly, in this manner one can construct one-dimensional coordinate systems for all real variables.

Now we shall begin the construction of a two-dimensional coordinate system. As shown in Fig. 3.2, again we place the x axis horizontally, but now we place the y axis at a right angle to the x axis with their origins coinciding. It is now possible to associate any point in the plane that these two lines determine, called the xy plane, with a *pair* of real numbers. These numbers are called, respectively, the x and the y coordinates of the point. Conventionally, they are enclosed by parentheses. The point with coordinates (3,2) is shown in Fig. 3.2. Clearly, it is that point in the plane where the vertical line through 3 on the x axis meets the horizontal line through 2 on the y axis. The general point in the plane has coordinates (x,y). It now seems obvious each such ordered pair of real numbers is geometrically associated with a point in the plane and also that each point in the plane has a pair of real numbers associated with it.

The graph of a function $y = f(x)$ is the set of all points in the xy plane just described whose first coordinate is x and whose second coordinate is y, that is, $f(x)$. Consider $y = 3x + 7$. Choose $x = 1$. Then $y = 10$, and the point P with coordinates (1,10) is a point of the graph of the function. Obviously, any value for x can be chosen, and the corresponding y easily calculated. The point (x,y) lying on the graph can then just as easily be plotted in the xy plane. It is a fact that the set of points (x,y) satisfying this equation lie on a straight

Fig. 3.1

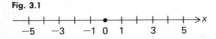

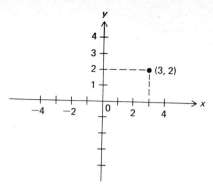

Fig. 3.2

line.   Plot a few more points and see if you can verify this.   You recall, though, that a straight line is determined if two distinct points which lie on it are prescribed.   It would appear, then, that the line itself could easily be drawn by simply plotting very carefully *two* points on it and then extending the line segment joining these two points indefinitely in both directions.   The points $(1,10)$ and $(-2,1)$ satisfy the equation and thus lie on the line.   The graph is shown in Fig. 3.3.   If you employ this two-point method for graphing straight lines, it would be best if the two points chosen are not too close to each other.

In general, the graph of the linear equation $y = ax + b$ is a straight line. a and b are constants.

Some lines have peculiar equations.   Consider the line with equation $y = 2$.   Clearly, whatever x one chooses, the corresponding y is 2.   Two points on the line are, for example, $(3,2)$ and $(-7,2)$.   Evidently, $y = 2$ is the equation

Fig. 3.3

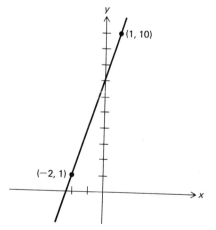

of a straight line parallel to the x axis and two units above it. Again, consider the equation $x = -3$. $(-3,4)$, $(-3,8)$, $(-3,5)$, etc., are all points of the line. All points of the plane with x coordinate -3 are points of the line. Clearly, $x = -3$ is the equation of a line parallel to the y axis and three units to the left of it. We leave it to you to generalize to the family of lines $y = a$ and to the family of lines $x = b$, where a and b are any constants.

Now one ought to be able to write the *equation* of a line if one knows two points on it. We state the following rule: If (x_1,y_1) and (x_2,y_2) are two distinct points on a straight line, then its equation is

$$y - y_1 = \frac{y_2 - y_1}{x_2 - x_1} (x - x_1) \tag{3.2}$$

or, equivalently,

$$y - y_2 = \frac{y_2 - y_1}{x_2 - x_1} (x - x_2) \tag{3.3}$$

Note that x and y are the *variables* here and that x_1, x_2, y_1, and y_2 are constants. The resulting equation is linear and hence represents a straight line. Using the first version, we now verify that both (x_1,y_1) and (x_2,y_2) lie on the line. We substitute x_1 for x, y_1 for y; then $0 = \frac{y_2 - y_1}{x_2 - x_1} (0) = 0$; hence, (x_1,y_1) lies on the line. Now we substitute x_2 for x and y_2 for y; then

$$y_2 - y_1 = \frac{y_2 - y_1}{x_2 - x_1} (x_2 - x_1)$$
$$= y_2 - y_1$$

Fig. 3.4

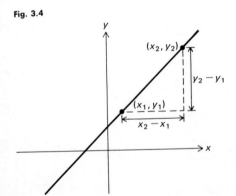

hence, (x_2, y_2) is on the line. We leave it to you to verify that (x_1, y_1) and (x_2, y_2) satisfy Eq. (3.3).

The quantity $(y_2 - y_1)/(x_2 - x_1)$ which appears in both Eqs. (3.2) and (3.3) is the *slope* of the line. It is usually denoted by m. Consider Fig. 3.4. It shows a straight line, and the points (x_1, y_1) and (x_2, y_2) are any two distinct points on it. The vertical distance $y_2 - y_1$ is sometimes called the *rise* of the line; the horizontal distance $x_2 - x_1$ is called the *run*. The ratio of the rise to the run is the slope of the line, that is, $m = (y_2 - y_1)/(x_2 - x_1)$. It seems evident that the slope of a line is constant. In fact, it can be shown that the only plane curve with its slope constant is the straight line.

Equation (3.2) can be written

$$y - y_1 = m(x - x_1) \tag{3.4}$$

This version of the equation of a straight line is called the *point-slope form* since it allows one to find the equation of the line if its slope m and a point (x_1, y_1) on the line are given.

Example. Find the equation of the straight line which passes through the points $(4,2)$ and $(-3,5)$.

Solution: Using Eq. (3.2) with $x_1 = 4$, $x_2 = -3$, $y_1 = 2$, $y_2 = 5$, we get

$$y - 2 = \frac{5 - 2}{-3 - 4}(x - 4)$$

$$y - 2 = \frac{-3}{7}(x - 4)$$

$$7y - 14 = -3x + 12$$

or, finally, $3x + 7y - 26 = 0$. The slope of this line is $-3/7$.

Example. Find the equation of the straight line whose slope is -4 and which passes through the point $(3,6)$.

Solution: Using (3.4), with $x_1 = 3$, $y_1 = 6$, $m = -4$, we get $y - 6 = -4(x - 3)$, $y - 6 = -4x + 12$, or $y = -4x + 18$.

We note that if m is positive, the line rises as it moves to the right; if m is negative, the line falls as it moves to the right; and if $m = 0$, the line is parallel to the x axis. The slope of a vertical line is not defined.

Equation (3.4) for the straight line can be rewritten as follows: $y - y_1 = m(x - x_1)$, $y = mx + (y_1 - mx_1)$, $y = mx + b$, where $b = y_1 - mx_1$. The new form for the equation of the line $y = mx + b$ is called the *slope-intercept form*. Clearly, m is its slope. But note that when $x = 0$, $y = b$. Thus, the point $(0,b)$ is on the line, and this is the point at which the line intersects the y axis. The number b is called the y intercept.

Example. Find the equation of the line whose slope is 3 and whose y intercept is -5.

Solution: Using the slope-intercept form with $m = 3$ and $b = -5$, we have $y = 3x - 5$.

Example. Find the slope of the line $3x + 4y = 8$.

Solution: We place the equation into the slope-intercept form. Thus, $4y = -3x + 8$, $y = -3x/4 + 2$. Therefore, $m = -3/4$, and the y intercept is 2.

EXERCISES

3.14 Find the equation of the line which passes through the points
(a) $(3,4)$ and $(-2,5)$
(b) $(-4,0)$ and $(5,6)$
(c) $(4,1)$ and $(-3,1)$
(d) $(0,0)$ and $(-6,2)$
(e) $(5,6)$ and $(5,0)$
(f) $(11,0)$ and $(0,4)$

3.15 Plot the lines with the following equations:

(a) $y = 3x + 4$
(b) $y = -4$
(c) $x + y = 3$
(d) $3x - 2y = 5$
(e) $x = 5$
(f) $y = -4x$
(g) $y = -3x + 4$
(h) $y = x$

3.16 Find the slope of each of the following lines:

(a) $y = x$
(b) $3x + 5y = 7$

(c) $x - y = 8$

(d) $\dfrac{x}{2} + \dfrac{y}{3} = \dfrac{4}{5}$

(e) $y = 5$

3.17 Find the equation of the line with the following properties:

(a) passes through $(0,0)$ and $(4,6)$;

(b) passes through $(-5,6)$ and has slope -4.

(c) has y intercept 4 and slope 6.

(d) passes through $(4,5)$ and is parallel to the line $2x - 4y = 9$.
Hint: Two lines are parallel if they have the same slope.

(e) passes through $(-3,-2)$ and has y intercept 3.

(f) has y intercept 5 and x intercept -4.

3.18 The line $y = mx + 7$ passes through the point $(6,7)$. What is its slope?

3.19 Two variables F and C are related linearly, that is, $F + bC = d$, where b and d are constants. If $F = 32$ when $C = 0$, and $F = 212$ when $C = 100$, find b and d.

3.20 The linear equation of Exercise 3.19 represents the relation between Fahrenheit and Celsius temperatures. (a) Use the result of this exercise to find the Celsius temperature C corresponding to the Fahrenheit temperature $F = 450°$. (b) Use the result of Exercise 3.19 to find the temperature at which $F = C$.

3.21 The position of a particle is given by $s = 3t + 8$, where s is in feet and t is in seconds. (a) Find s when $t = 7$. (b) Find t when $s = 29$.

3.4 SIMULTANEOUS LINEAR EQUATIONS

We suppose you have already learned how to solve two simultaneous linear equations like

$$\begin{cases} x + 2y = 5 \\ 2x - y = 7 \end{cases} \qquad (3.5)$$

So, recall that a suitable method is called the *method of elimination*. First, we eliminate the variable x by multiplying the first equation of (3.5) by -2, then adding the resulting equation to the second equation of (3.5). Thus,

$$-2x - 4y = -10$$
$$\underline{2x - y = 7}$$
$$- 5y = -3$$
$$y = \frac{3}{5}$$

Then, using the first equation of (3.5), $x = 5 - 2y$. Substituting $y = {}^3/_5$, we have $x = 5 - {}^6/_5 = {}^{19}/_5$. One easily verifies that substituting $x = {}^{19}/_5$, $y = {}^3/_5$ into (3.5) satisfies both equations; i.e., they form a solution set.

From our discussion in Sec. 3.3 we note that each of the equations of (3.5) can be represented geometrically as a straight line. To solve the equations algebraically corresponds geometrically to finding the coordinates of the point at which the two lines intersect. Figure 3.5 shows these lines and their point of intersection.

There are occasions, of course, when the two lines do not intersect. If the lines are distinct and parallel to each other, they will have no point of intersection. Consider

$$\begin{cases} x + 4y = 7 \\ 2x + 8y = 17 \end{cases}$$

If we follow the usual procedure and try to eliminate x from the two equations by multiplying the first equation through by -2 and adding, we have

Fig. 3.5

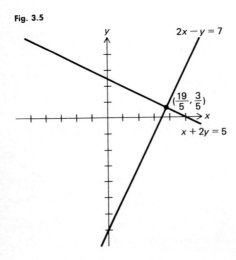

$$-2x - 8y = -14$$
$$\underline{2x + 8y = 17}$$
$$0 = 3$$

Both variables are eliminated, and the false result $0 = 3$ ensues. As you can check, the two lines represented by the equations are parallel.

It is possible that the two linear equations actually represent the *same* line. Consider

$$\begin{cases} 2x - 5y = 7 \\ 4x - 10y = 14 \end{cases}$$

Elimination of x now leads to the true result $0 = 0$. Closer inspection reveals that the second equation is merely the first one multiplied through by 2 and thus represents geometrically the same set of points as the first equation. If one *must* consider this simultaneous set as representing two lines, even though they are not distinct, then *every* point lying on one satisfies the other equation. The solution set is thus the entire (infinite) set of points satisfying either equation. Some of the infinitely many solutions are $(0,-7/5)$, $(2,-3/5)$, and $(7/2,0)$.

The algebraic procedure followed above can be used for the solution of three simultaneous linear equations in three unknown variables. We shall not present the geometric significance in detail except to say that the three equations now represent planes in three-dimensional space and we are trying to find the point (with three coordinates) in which the three planes intersect. Consider the following:

$$\begin{cases} x + y + 3z = 5 & \quad (3.6a) \\ x + 2y = 1 & \quad (3.6b) \\ 2x - z = 4 & \quad (3.6c) \end{cases}$$

We eliminate z from (3.6a) and (3.6c) by multiplying (3.6c) through by 3 and adding the result to (3.6a). Thus,

$$x + y + 3z = 5$$
$$\underline{6x - 3z = 12}$$
$$7x + y = 17 \qquad (3.6d)$$

Now combining (3.6d) with (3.6b) which already contains only x and y, we have

$$7x + y = 17 \qquad (3.6d)$$
$$x + 2y = 1 \qquad (3.6b)$$

Multiplying (3.6d) through by -2 gives

$$-14x - 2y = -34$$
$$\underline{x + 2y = 1}$$
$$-13x \qquad = -33$$
$$x = \frac{33}{13}$$

Using (3.6b), which is $x + 2y = 1$, we have $2y = 1 - x = 1 - \frac{33}{13} = (13 - 33)/13 = -\frac{20}{13}$; therefore, $y = -\frac{10}{13}$. Now using (3.6c), which is $2x - z = 4$, gives $z = 2x - 4 = 2(\frac{33}{13}) - 4 = \frac{66}{13} - \frac{52}{13} = \frac{14}{13}$; therefore, $z = \frac{14}{13}$. Thus, the solution set is $x = \frac{33}{13}$, $y = -\frac{10}{13}$, $z = \frac{14}{13}$.

EXERCISES

3.22 Solve the following sets of simultaneous linear equations:

(a) $\begin{cases} x + 3y = 8 \\ 3x - y = 7 \end{cases}$

(b) $\begin{cases} 2x + 5y = 12 \\ x - 7y = -2 \end{cases}$

(c) $\begin{cases} a + b = 3 \\ 2a + b = 13 \end{cases}$

(d) $\begin{cases} 3x + 4y = 16 \\ 6x + 8y = 32 \end{cases}$

(e) $\begin{cases} x - y + z = 5 \\ 2x - y - z = 8 \\ x \quad + z = -3 \end{cases}$

(f) $\begin{cases} a + 2b + 3c = 25 \\ a - 3b - 4c = -36 \\ 4a - 5b + 7c = 19 \end{cases}$

(g) $\begin{cases} \dfrac{x}{4} + \dfrac{y}{5} = \dfrac{6}{7} \\ \dfrac{x}{6} + \dfrac{y}{7} = \dfrac{8}{9} \end{cases}$

(h) $\begin{cases} 0.12u + 0.26v = -0.45 \\ 0.03v + 0.05u = -0.04 \end{cases}$

(i) $\begin{cases} 3x + 4y + 5z = 6 \\ 2y - 3z = -4 \\ 2z = 7 \end{cases}$

3.23 Solve the following sets of simultaneous linear equations if all the

Fig. 3.6

coefficients and the variables are in the given base. All the operations required to solve the equations are to be done *using that base*.

(a) Base 2: $10x + y = 101$
$$x - y = 100$$
(b) Base 4: $10x - y = 21$
$$3x - 2y = 10$$
(c) Base 9: $6x + 5y = 38$
$$18x - 16y = 110$$
(d) Base 16: $20x - y = 3D$
$$11x + Dy = 49$$

3.5 LINEAR INEQUALITIES IN ONE VARIABLE. ABSOLUTE VALUE

Consider the coordinate system of Fig. 3.6. Note that two points labeled a and b are marked on it. Clearly, b is greater than a because it is farther to the right on the coordinate system than a is. We write this as $b > a$ (read "b is greater than a"). At the same time, of course, a is less than b since it is farther to the left than b is. We write this as $a < b$ (read "a is less than b"). Note that $a < b$ is just another version of $b > a$ read (and written) from right to left. The two statements $b > a$ and $a < b$ are examples of *inequalities*. Here are some examples of inequalities that are evidently true:

$$-4 < 2 \qquad 5 > -8 \qquad 7.3 < 12.6$$

It seems fairly evident that if a and b are *any* two real numbers, exactly one of the following must hold: $a < b$, $a > b$, or $a = b$.

We now proceed to geometrically interpret some linear inequalities in one variable.

Example. Show the set of all numbers x such that $x > 2$.

Solution: Draw the one-dimensional coordinate system for the variable x and mark on it all numbers that are greater than 2:

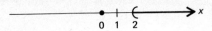

Fig. 3.7

This is evidently the set of all numbers to the right of 2; that this set does not include 2 itself is designated on Fig. 3.7 by the (, the opening parenthesis, at 2.

Example. Show the set of all x such that $x \leq 3$. We read this inequality "x is less than or equal to 3." The set includes 3 and is shown in Fig. 3.8. The symbol] indicates that $x = 3$ is to be included in the set of values.

We read $2 < x < 5$ as "x is greater than 2 and (at the same time) less than 5." Figure 3.9 shows this set of values of x. This is called an *open in-terval*—it contains all values of x between 2 and 5 but does not include the end points 2 and 5 themselves. Figure 3.10 shows geometrically the set of values of x satisfying $-3 \leq x < 2$. This is another interval but it contains the left end point -3; it is called *half-closed*. Another notation for the set of values of x between 2 and 5 is (2,5), which is read "the open interval from 2 to 5." Context is used so that this notation is not confused with the same notation used for coordinates of a point. If an interval contains its left end but not the right end, it is written with a square bracket on the left, as in [3,7).

Finally, the notation [2,12] is used to represent the *closed* interval (all x such that x is greater than or equal to 2 and is less than or equal to 12). If an interval extends indefinitely to the right, we write (a,∞), meaning all x such that x is greater than a. We also write such expressions as $(-\infty,5]$ for $-\infty < x \leq 5$, meaning all x which are less than or equal to 5. We never close the interval at ∞ since ∞ is not a number but is used to indicate indefinite ex-tension to the left or right. It is the *infinity* symbol.

The notion of *absolute value* is of some importance in many applied problems. We write, for example, $|a|$ and read "the absolute value of a." The absolute value of a is simply a itself if a is zero or positive, i.e., if $a \geq 0$.

Fig. 3.8

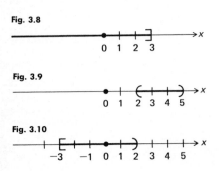

Fig. 3.9

Fig. 3.10

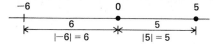

Fig. 3.11

It is $-a$ if a is negative, i.e., if $a < 0$. Thus, $|0| = 0, |+5| = 5, |-7| = -(-7) = 7$. The result is, as you can easily see, that the absolute value of any real number is just its *magnitude* and is always positive or zero. To interpret $|a|$ geometrically, we draw a one-dimensional coordinate system and place a on it. $|a|$ is the distance, always ≥ 0, that the point a is from the origin. Figure 3.11 shows the idea.

Consider the equation $|x - 2| = 3$. What could x be? Sheer guessing shows that x could be 5 or -1. There is a way to proceed algebraically to find x. If the absolute value of "something" is 3, then the "something" must be 3 or -3. Thus, in this case, the "something" $x - 2$ must either be 3 or -3. Thus, $x - 2 = 3$ or $x - 2 = -3$. Solving each of these, we arrive at $x = 5$ or $x = -1$.

Example. Solve for x: $|7 - 3x| = 6$.

$$7 - 3x = 6 \qquad 3x = 1 \qquad x = \frac{1}{3}$$
$$7 - 3x = -6 \qquad 3x = 13 \qquad x = \frac{13}{3}$$

Thus, $x = \frac{1}{3}$ or $\frac{13}{3}$. Either of these values of x, and none other, satisfies the equation $|7 - 3x| = 6$.

Inequalities can involve absolute value. Consider the inequality $|x| < 3$. Clearly, x could equal 2, -1, -2.5, 1.7, etc. But a check reveals that x cannot be greater than 3 or less than -3. Thus, $|x| < 3$ is exactly the same set of values of x as $-3 < x < 3$, that is, the open interval $(-3,3)$. In general, if the absolute value of an expression is less than a, then the expression itself must be between $-a$ and a. **Example.** $|7 - 5x| < 8$ is equivalent to $-8 < 7 - 5x < 8$.

Consider the inequality $|x - 3| > 5$. Values of x like 12, -7, 347, etc., satisfy this inequality. A check reveals that for this inequality to hold, the quantity $x - 3$ must be either greater than 5 or less than -5. This means that $x > 8$ or $x < -2$. This set of values of x is not an interval on the x axis, as Fig. 3.12 reveals.

Fig. 3.12

It is possible to solve inequalities in one variable in *almost* exactly the same way one solves linear equations.

Example. Solve $2x + 3 = -8$ for x. Clearly, $2x = -11$ (by subtracting 3 from both sides) and $x = -11/2$ (by dividing both sides of the equation by 2). x has exactly *one* value.

Example. Solve $2x + 3 < -8$ for x. Adding -3 to both sides of the inequality, we have $2x < -11$. Then dividing both sides by 2, $x < -11/2$. In other words, if *any* value of x less than $-11/2$ (like -7) is chosen, the expression $2x + 3$ will have value less than -8.

We have the following first rule: Adding or subtracting a number from both sides of an inequality does not alter in any way the meaning and mathematical content of the inequality. Special care must be taken, however, when both sides of an inequality are multiplied or divided by a number (not zero). Consider the inequality $4 - x < 3$. Suppose we subtract 4 from both sides. We now have $-x < -1$. So far so good. But we want to isolate x on one side. Two methods now present themselves. Add x to both sides, $0 < -1 + x$, now add 1 to both sides, $1 < x$, with the result $x > 1$. Let us now reconsider $-x < -1$. *Multiplying* both sides by -1, we must at the same time *reverse* the inequality in order to arrive at the correct result, and again end with $x > 1$.

In general, the sense of the inequality must be reversed when both sides of the inequality are multiplied or divided by a *negative* number. We have this rule, then: To solve a linear inequality for x, follow normal algebraic procedures to isolate x on one side of the inequality but if this procedure involves multiplying or dividing both sides by a negative number, reverse the inequality at this step.

Example. Solve $4 - 3x < 10$ for x. We have

$-3x < 6$ (by subtracting 4 from both sides)
$x > -2$ (by dividing both sides by -3 and reversing the inequality)

One can solve combined inequalities for x, too.

Example. Solve $8 \leqslant 41 - 5x < 76$ for x. We have

$-33 \leqslant -5x < 35$ (adding -41 to both sides)
$\dfrac{33}{5} \geqslant x > -7$ (by dividing all parts by -5 and reversing both inequalities)

Fig. 3.13

We can solve *simultaneous* linear inequalities in one variable using a semigraphical method.

Find all x satisfying *both* of the following inequalities: $3 < x < 8$, $-1 < 3x - 7 < 11$. Solving the second inequality for x, we have $2 < x < 6$. We then plot on the same x coordinate system the sets of x satisfying each of the inequalities $3 < x < 8$ and $2 < x < 6$. If x is to satisfy both inequalities, clearly we should choose x in the part where the two regions overlap. Thus, $3 < x < 6$ is the solution, as shown in Fig. 3.13.

Example. Solve the simultaneous inequalities

$$\begin{cases} -2 \leqslant x < 4 \\ -2 \leqslant 3 - x < 2 \end{cases}$$

We leave it to you to show that the second inequality is equivalent to $1 < x \leqslant 5$. Figure 3.14 shows the plot of these inequalities. Thus, the solution is $1 < x < 4$.

EXERCISES

3.24 Using the method illustrated in Figs. 3.7 and 3.8, show geometrically the set of all x such that

(a) $x \geqslant 4$
(b) $-2 \leqslant x < 4$
(c) $-\infty < x \leqslant -3$
(d) $2 \leqslant x < 7$
(e) $3 \leqslant x < \infty$
(f) $2 < x < \dfrac{7}{2}$

3.25 Express each of the intervals of Exercise 3.24 using the notation described after Fig. 3.10.

Fig. 3.14

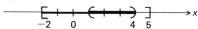

3.26 Solve for x:

(a) $|x + 1| = 5$
(b) $|4 - 5x| = 7$
(c) $|2x - 7| = 0$

3.27 Solve the following linear inequalities for the variable contained:

(a) $3y - 7 > 9$
(b) $2 - 7x < 12$
(c) $\dfrac{x - 7}{3} \geqslant -5$
(d) $6 - 3p > 0$
(e) $-8 < q + 7 < -3$
(f) $0 \leqslant 2 - 5x \leqslant 12$
(g) $\dfrac{3}{4} \leqslant \dfrac{6x - 7}{8} < \dfrac{9}{2}$
(h) $1 - x < 3x + 8$
(i) $-7 < 1 - \frac{1}{2}w \leqslant -3$

3.28 Use the semigraphical method of the text to solve the following simultaneous linear inequalities:

(a) $\begin{cases} -3 < x < 5 \\ 0 < x + 2 < 9 \end{cases}$

(b) $\begin{cases} 3 \leqslant 2x - 7 \leqslant 5 \\ 3 < x + 3 \leqslant 8 \end{cases}$

(c) $\begin{cases} 2 \leqslant x < \infty \\ 0 < x + 3 < 10 \end{cases}$

(d) $\begin{cases} -5 \leqslant 2x + 1 < 15 \\ -2 \leqslant 1 - x < 6 \end{cases}$

3.29 Solve the following inequalities for x:

(a) $|2x| < 6$
(b) $|3 - x| \leqslant 5$
(c) $\left| \dfrac{x}{4} - \dfrac{1}{5} \right| < 2$
(d) $|x + 7| > 3$
(e) $|7 - 3x| > 4$

3.6 LINEAR INEQUALITIES IN TWO VARIABLES

We complete this section on inequalities with a discussion of the graphing of linear inequalities in *two* variables. Consider $2x + 3y > 7$. Clearly, a great number of points (x,y) in the xy plane satisfy this inequality, like $(1,2)$, $(2,2)$, $(-1,5)$, and $(0,3)$. In fact, as we shall soon see, there is an entire half-plane of points that satisfy it. Here is how we proceed to find the part of the xy plane all of whose points satisfy $2x + 3y > 7$. First, consider for a moment the set of points that satisfy the *equation* $2x + 3y = 7$, geometrically a straight line. Figure 3.15 shows this line plotted in the xy plane. This line divides the plane into two half-planes, the half-plane above the line and the half-plane below the line. In one of these half-planes the expression $2x + 3y$ is always greater than 7 for any point (x,y) in it and in the other half-plane $2x + 3y$ always has a value less than 7. To discover which is which, we just select at random a point known to lie in one of the half-planes (not on the line). Suppose we select $(5,0)$. By simple substitution into $2x + 3y$ (let $x = 5$, $y = 0$), we find that it has the value $10 + 0 = 10$. Since $10 > 7$, we conclude that $(5,0)$ belongs to the half-plane defined by $2x + 3y > 7$ since it satisfies the inequality. Since $(5,0)$ is clearly in the upper half-plane, we then conclude that all the points above the line will satisfy the inequality $2x + 3y > 7$ and that this half-plane is in fact the set of all points which satisfy the inequality. We have noted this region on the graph of Fig. 3.15. We now reason that all the points below the line satisfy the inequality $2x + 3y < 7$. [Just to be sure, though, we note that $(0,0)$ is in this region and it obviously satisfies the inequality $2x + 3y < 7$ since $0 < 7$.] We see now that actually only one point need be tested, and the character of the entire plane with reference to the inequality can be determined.

Fig. 3.15

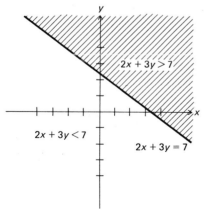

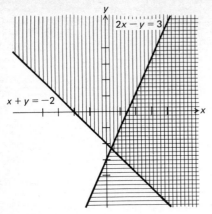

$2x - y = 3$

$x + y = -2$

Fig. 3.16

Now we are ready to consider simultaneous linear inequalities in two variables. Suppose we are to find all points in the xy plane which satisfy both $2x - y > 3$ and $x + y > -2$. As shown in Fig. 3.16, we first plot the two lines $2x - y = 3$ and $x + y = -2$. Then, using the method just described, we find and shade *horizontally* the half-plane corresponding to $2x - y > 3$. It is clearly the half-plane to the right of the line [for example, $(5,0)$ lies in this half-plane and it satisfies the inequality $2x - y > 3$ since $10 > 3$]. Next we find and shade *vertically* the half-plane corresponding to the inequality $x + y > -2$. It is clearly the half-plane above the line since $(0,0)$ lies in this half-plane and it satisfies the inequality since $0 > -2$. Finally, we reason that the set of points satisfying *both* the inequalities is simply the intersection of the two shaded half-planes, that is, the parts they have in common. The double-shaded region that appears in Fig. 3.16 (this does not include the points on the lines themselves) is the region we have been seeking.

It should now be clear to you how you could geometrically represent the region in the xy plane all points of which satisfy *any number* of simultaneous linear inequalities in two variables x and y. Here is an example of three such inequalities.

Example. Find and sketch the region determined by the inequalities $x + y < 2$, $x - y > -5$, and $2x + 4y > 5$. The three lines $x + y = 2$, $x - y = -5$, and $2x + 4y = 5$ are shown in Fig. 3.17. We leave it to you to show that the half-plane below the line $x + y = 2$, the half-plane above the line $2x + 4y = 5$, and the half-plane to the right of the line $x - y = -5$ are the proper half-planes and that the shaded area shown in Fig. 3.17 is the region required.

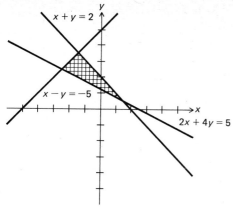

$x + y = 2$

$x - y = -5$

$2x + 4y = 5$

Fig. 3.17

3.30 Plot and shade the half-plane determined by each of the following inequalities:

   (a) $x + y > -3$
   (b) $3x + 4y < 12$
   (c) $x < -5$
   (d) $y > 3$
   (e) $2x - 3y < 18$

3.31 Plot and shade the region determined by each of the following sets of inequalities:

   (a) $\begin{cases} 2y > x \\ 3x + 5y > 15 \end{cases}$

   (b) $\begin{cases} x > 2 \\ x + y < 4 \end{cases}$

   (c) $\begin{cases} 3x - 2y > 12 \\ x + 7y > 14 \end{cases}$

   (d) $\begin{cases} 2x + 5y < 7 \\ x - y < 4 \end{cases}$

   (e) $\begin{cases} 5x + 6y > 60 \\ 4x + 5y < 56 \end{cases}$

3.32 Plot and shade properly the area determined by the following sets of inequalities:

(a) $\begin{cases} -2x + 3y > -6 \\ 4x + 3y > 12 \\ 6x - 3y > -18 \end{cases}$

(b) $\begin{cases} -2x + 3y > -6 \\ 4x + 3y < 12 \\ 6x - 3y > -18 \end{cases}$

(c) $\begin{cases} y > 4 \\ y > x \\ x > -2 \end{cases}$

(d) $\begin{cases} 2x - 4y < -8 \\ 3x + 2y > 6 \\ 3x + 4y < 12 \end{cases}$

(e) $\begin{cases} y < 4 \\ y < x \\ 2x + y < 12 \\ -2x + 5y > -10 \end{cases}$

(f) $\begin{cases} -x + 2y > -2 \\ x + y < 4 \\ x + y > -2 \\ 4x - 3y > -12 \end{cases}$

3.7 LINEAR PROGRAMMING

Linear programming is a subject that we can pursue only through its elementary ideas and applications but which plays an extremely important role in many decision problems of business and industry. It is often concerned with optimum allocation of certain resources to meet certain commitments and objectives but subject to definite restrictions. Interestingly enough, as you could guess from the word *linear* in its name, it is concerned mathematically entirely with such problems as the solution of simultaneous sets of equations and inequalities; so, at least to start, it should not be beyond our understanding.

The main idea is that a linear expression (which could be representing such a thing as profit) is to be maximized when the variables it contains are not specifically given but are to be determined subject to restrictions in the form of linear equations and/or linear inequalities that these variables must satisfy. Another way to say this is that the given linear expression, often called the *object* function, is to be *optimized* by choosing values for the variables that it contains but that these variables are themselves subject to severe restrictions that are also specified. Although we are not yet mathematically prepared to read its text completely, the book "Linear Programming" by Saul I. Gass, McGraw-Hill, New York, 1964, in its introductory chapter describes many different practical problems that the processes of linear programming can solve. Some of these are the transportation problem, the diet problem, the allocation of contract awards, personnel allocations, production scheduling, and inventory control. Applications are made to such industries as agriculture, chemicals, commercial airlines, and petroleum.

One must keep in mind that because of the type of quantities, such as amounts of material and number of personnel, which the variables represent, the solution values of the variables must always be nonnegative. Let us proceed, as usual, by presenting an example. Suppose we have the object function $z = 5x + 11y$ and we wish to find the maximum value of z if the variables x and y are restricted by the inequalities $x + y \leqslant 5$ and $3x - 2y \leqslant 6$. That is, we must find $x \geqslant 0$ and $y \geqslant 0$, subject to these restrictive inequalities, that will make the linear expression $5x + 11y$ as large as possible. To proceed, let us plot the lines $x + y = 5$ and $3x - 2y = 6$, then find the region R corresponding to the set of inequalities

$$x \geqslant 0$$
$$y \geqslant 0$$
$$x + y \leqslant 5$$
$$3x - 2y \leqslant 6$$

This will at least display the region in the xy plane we shall be restricted to. Figure 3.18 shows this plot and the region R.

To find the region R, we have, of course, followed exactly the method

Fig. 3.18

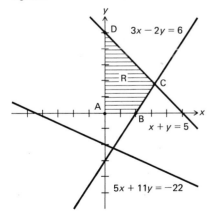

discussed in the last section. The region R *including* all its edges is the set of all pairs (x,y) satisfying all four restrictive inequalities. Now we select, if we can, a point (x,y) in R which will make the value of z, that is, the object function $5x + 11y$, as large as possible. The corners of R have coordinates $A(0,0)$, $B(2,0)$, $C(^{16}/_5,^9/_5)$, and $D(0,5)$. These points are found by solving simultaneously the pairs of equations of lines that intersect at the point. It turns out that it is quite easy to find this interesting point (x,y) in R.

We can regard $z = 5x + 11y$ as representing a *set* of straight lines, each member of which we get by assigning a definite value to z. Such lines as $5x + 11y = 1$, $5x + 11y = 2$, $5x + 11y = 7$, and $5x + 11y = 47$ belong to this set. We note that each line of the set is parallel to every other line of the set, since they all have the same slope. On Fig. 3.18 we have drawn one such line, $5x + 11y = -22$. We now move this line upward parallel to itself, and we note that this simply amounts to changing the z in $5x + 11y = z$. The first vertex of R that the (moving) line will pass through is $A(0,0)$. Then $5x + 11y = 0$, since $(0,0)$ satisfies this equation. So z now has value 0, considerably greater than the value -22 it had originally. As we move the line higher, the value of $5x + 11y$, that is z, will increase since the points in R have positive coordinates. Soon the line will pass through $B(2,0)$. Now $5x + 11y$ has value 10, and z has value 10. We note while pausing here that all the other points in R now lying on the line will cause the linear function $5x + 11y$ to have the value 10. Again we move higher until the line passes through $C(^{16}/_5,^9/_5)$. Now $z = 5(^{16}/_5) + 11(^9/_5) = 16 + ^{99}/_5 = 33.8$, a still greater value for z. We now have a pretty good value for z but we hope to push on and again move the line parallel to itself and higher. We finally pass through $D(0,5)$ at which z has value $5(0) + 11(5) = 55$. This is clearly the maximum value $z = 5x + 11y$ could have *for any point of* R. Thus, z is maximum if $x = 0$, $y = 5$, and its maximum value is 55.

The result of all this moving of the line $z = 5x + 11y$ can be stated in the following rule: Evaluate $z = 5x + 11y$ at each of the *corners* of R. The largest value of z thus obtained is its maximum value of z on R; the minimum value of z thus obtained is the minimum value of z on R. Again recall that staying in R or on its edge means that we are maintaining the restrictions caused by the imposition of the given inequalities.

A simple practical interpretation of the problem we have just solved could be as follows. A manufacturer produces two items A and B. Let x be the number of items A produced each week and y be the number of items B produced each week. Because of manufacturing and other physical restrictions, the total number of A and B items must be less than 5, that is, $x + y \leqslant 5$. Suppose also that because of mysterious circumstances it is known that three times the number of A items less two times the number of B items is less than or equal to 6. Now suppose that each A item sells for $5 and each B item sells for \$11. Thus, the income is $5x + 11y$ (in dollars). Call this z. The problem can then be stated as "Find how many items A and how many items B should be produced each week in order that the income be maximized." As we have seen, it would be better if the manufacturer abandon item A since our analysis shows that he should produce no (zero) items A and five items B each week. His income will then be maximum at \$55.

This simple example, possible for us to do with the mathematical tools we have at hand, has given us a beginning notion of the extremely important applications of linear programming.

EXERCISES

3.33 Maximize $z = 10x + 7y$ under the conditions

$$\begin{cases} 8x + 5y \leqslant 40 \\ 3x - 2y \geqslant 6 \\ 4x - 9y \leqslant 36 \end{cases}$$

Recall that $x \geqslant 0$ and $y \geqslant 0$. Sketch the region R as discussed in the text.

3.34 Find the maximum value of $z = 42x + 17y + 16$ under the conditions $x \leqslant 7, 6y - 5x \leqslant 30$. Recall that $x \geqslant 0$ and $y \geqslant 0$.

3.35 Given the set of restrictions

$$\begin{cases} x - 4y \leqslant 8 \\ x + 3y \geqslant 6 \\ x - 4y \geqslant -8 \\ x \leqslant 12 \end{cases}$$

and also $x \geqslant 0, y \geqslant 0$,

(a) minimize $z = x + 2y + 7$
(b) maximize $z = 3x - 11y$

3.36 A manufacturer produces two items A and B. Let x be the number of items A produced each week and let y be the number of items B produced each week. Suppose his profit is \$25 for each item A produced and \$18 for each item B produced. To manufacture these items, three resource materials C, D, and E are used. The manufacturer has available each week a maximum of 24 of material C, 20 of material D, and 27 of material D. To produce one item A

requires eight material items C and five items D. To produce one item B requires six material items C, four items D, and nine items E. Determine the number of items A and B he should produce each week in order to maximize his profit. What is this maximum profit?

3.8 FITTING DATA TO A LINE. SUMMATION

The final application of linear functions we shall present in this chapter is curvefitting to a line. Suppose we are given a set of points, no matter how many but more than two points. For example, the data set could be

x	1	2	3	4	6
y	2	3	5	8	9

Let us plot these points $(1,2)$, $(2,3)$, $(3,5)$, $(4,8)$, $(6,9)$. Clearly the points do not lie on a straight line. We do not know any function $y = f(x)$ whose graph contains all the points — at least any simple function. It might be possible, though, to find a straight line L which even if it cannot possibly pass through all the points will pass "reasonably" close to the points. Such a line can be sketched by "eyeballing" as we have done in Fig. 3.19. Notice that this first approximation does not go through any of the points. We wonder whether there might not be an analytic way to determine what might be called the *best-fitting line*. There are many ways that people use to define what is meant by best fitting. Perhaps the most often used notion is called the method of *least squares*. It is used to fit data to many different kinds of curves, and

Fig. 3.19

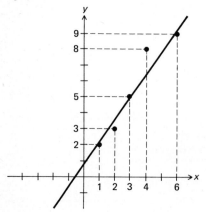

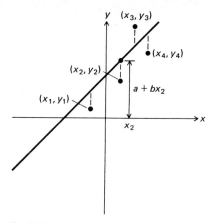

Fig. 3.20

we shall be using it again in a later chapter. For now we present the method as it applies to fitting data to a straight line. Let us generalize the process so that it will be necessary to use general symbols for the data, the x's and the corresponding y's. Thus, the data will be represented by (x_i, y_i), that is, by (x_1, y_1), (x_2, y_2), (x_3, y_3), . . . , (x_n, y_n), n points in all. On the graph in Fig. 3.20 we have shown some of these points.

We have also shown a line, whose properties are yet to be stated, which we want to be the best-fitting line. Now we want the line to pass relatively near if not go through the points, and it is this closeness that we must make precise. Suppose we let the equation of the line be $y = a + bx$. The points *on the line* which lie directly above (or below) the given data points have coordinates $(x_1, a + bx_1)$, $(x_2, a + bx_2)$, . . . , $(x_n, a + bx_n)$, since the y coordinate of the point corresponding to x_i is clearly $a + bx_i$ using the equation of the line. We look at the *vertical* deviations of the data points from the corresponding point on the line. This deviation is simply the difference in the y coordinates of the two points. Thus, the first point deviates $a + bx_1 - y_1$, the second point deviates $a + bx_2 - y_2$, and the third point deviates $y_3 - (a + bx_3)$, etc. Now we come to the central notion of the method of least squares. We shall impose the condition that the *sum of the squares* of each deviation shall be minimum. The best-fitting line is that line for which this sum is minimum. It is from the squaring and minimum ideas that the method of *least* squares takes its name. Thus we want the sum S, expressed in the equation

$$S = (a + bx_1 - y_1)^2 + (a + bx_2 - y_2)^2 + \cdots + (a + bx_n - y_n)^2$$

to be as small as possible. In this expression for S, all the x_i and y_i are known, being the data, so we must choose a and b in such a way as to minimize S. Once we have determined them, we insert a and b into $y = a + bx$ and have the equation of the best-fitting line. By use of processes of the differential calculus, it can be shown that the a and b we want must satisfy the following set of *linear* equations.

$$na + \left(\sum_{i=1}^{n} x_i \right) b = \sum_{i=1}^{n} y_i$$

$$\left(\sum_{i=1}^{n} x_i \right) a + \left(\sum_{i=1}^{n} x_i^2 \right) b = \left(\sum_{i=1}^{n} x_i y_i \right) \tag{3.7}$$

Before we can proceed, the notations used in Eq. (3.7) require some explanation. Consider

$$\sum_{i=1}^{n} x_i$$

The symbol Σ is the (capital) Greek letter sigma, and in mathematics it stands for *sum*. The x_i are of course the one-dimensional array of the given x values of the points of the data, that is, $x_1, x_2, x_3, x_4, \ldots, x_n$. The notation

$$\sum_{i=1}^{n} x_i$$

means that we are to sum the x_i as i (called the *index*) runs through integers from 1 to n, inclusive. Thus,

$$\sum_{i=1}^{n} x_i$$

simply means to add up the x's of the data. That is,

$$\sum_{i=1}^{n} x_i = x_1 + x_2 + x_3 + \cdots + x_n$$

Similarly,

$$\sum_{i=1}^{n} y_i = y_1 + y_2 + y_3 + \cdots + y_n$$

and

$$\sum_{i=1}^{n} x_i^2 = x_1^2 + x_2^2 + x_3^2 + \cdots + x_n^2$$

that is, the sum of the *squares* of each of the *x*'s of the data. Finally,

$$\sum_{i=1}^{n} x_i y_i = x_1 y_1 + x_2 y_2 + \cdots + x_n y_n$$

Before we turn back to our problem of fitting a line to data, a few more remarks and examples on this summation notation in general might be appropriate. Consider

$$\sum_{j=1}^{5} (2j + 3)$$

This means that we are to evaluate the function $f(j) = 2j + 3$ for j running from 1 to 5 and sum these values. Thus

$$\sum_{j=1}^{5} (2j + 3) = [2(1) + 3] + [2(2) + 3] + [2(3) + 3] + [2(4) + 3] + [2(5) + 3]$$
$$= 5 + 7 + 9 + 11 + 13 = 45$$

Similarly,

$$\sum_{k=2}^{4} g(k) = g(2) + g(3) + g(4)$$

whatever $g(k)$ is. Interestingly,

$$\sum_{i=1}^{10} i = 1 + 2 + 3 + 4 + 5 + 6 + 7 + 8 + 9 + 10$$
$$= 55$$

the sum of the first 10 integers (greater than 0). Finally, take note of this example:

$$\sum_{i=4}^{8} 7 = 7 + 7 + 7 + 7 + 7$$
$$= 35$$

Now that we know the meaning of the summation symbols used in Eq. (3.7), we return to our line-fitting example. Let us compute the various sums required by filling in the following table:

i	x_i	y_i	x_i^2	x_iy_i
1	1	2	1	2
2	2	3	4	6
3	3	5	9	15
4	4	8	16	32
5	6	9	36	54
Σ	16	27	66	109

Using (3.7), we now have

$$\begin{cases} 5a + 16b = 27 \\ 16a + 66b = 109 \end{cases}$$

Multiplying the first equation through by 16 and the second through by -5, we have

$$\begin{cases} 80a + 256b = 432 \\ -80a - 330b = -545 \end{cases}$$

Adding, we arrive at $-74b = -113$, $b = {}^{113}/_{74}$. Then using the first equation again, we have $5a = 27 - 16b = 27 - 16({}^{113}/_{74}) = 27 - 113({}^{8}/_{37}) = 27 - {}^{904}/_{37} = (999 - 904)/37 = {}^{95}/_{37}$. Thus, $a = {}^{19}/_{37}$.

The equation of the line is therefore $y = \dfrac{19}{37} + \dfrac{113}{74}\, x$. We leave it to you to check that none of the given points lies on this line. As a small check, let us find the point on the line whose x coordinate is 4. Substituting $x = 4$ into the equation, we have $y = {}^{19}/_{37} + 113({}^{2}/_{37}) = {}^{245}/_{37} \doteq 6.62$. Note that the data call for y to be 8 when $x = 4$.

Some final remarks may be helpful here. If the given data are collinear (that is, if they actually lie on a straight line), the method of least squares will give that line as the best fitting. Note that a line can be fitted to any data, no matter how many points it contains and no matter whether the points, when plotted, even appear to be vaguely collinear. In the latter case, it seems evident that the choice of the straight line would be highly inappropriate. Some other standard curve could undoubtedly be found which would fit the data much better than a straight line. As mentioned before, in later chapters we shall have more to say about the selection of other curves for fitting use and how to determine them by the method of least squares.

3.37 Show that the three points $(0,7)$, $(1,15)$, and $(3,31)$ lie on the line $y = 7 + 8x$. Now use the method of least squares to determine the best-fitting line for the given data.

3.38 Use the method of least squares to fit a line to the following data:

x	−1	0	2	3
y	2	1	−5	−8

3.39 Plot the data points given in the following table:

x	−2	0	2	4	6	7	8	10
y	2	3	5	5	5	7	7	11

Use the method of least squares to determine the best-fitting line. Plot the line *on the same graph*.

3.40 Fit the following to a straight line using the method of least squares:

x	1	2	3	4	5	6	7
y	−2	−1	−1	0	1	3	5

Recall the sum of squares called S in the text. Use the equation of the line you have just found to compute the value of S (that is, the minimum value that the sum of the squares of the deviations will have).

3.41 The voltage V across a certain resistor is measured as the current I through it is held constant and the resistance R is varied. The data resulting are as follows (R is in ohms, V in volts):

R	20	40	60	80	100
V	9.82	20.21	29.42	40.48	49.72

Use the method of least squares to fit these data to the linear equation $V = a + bR$. Recall that I was held constant. Suppose I is known to be ½ amp. Compare and discuss your resulting linear equation with Ohm's law, $V = IR$.

4

Introduction to Symbolic
Programming Languages, Fortran

4.1 INTRODUCTORY REMARKS

Every digital computer combines basic components called the *input, storage, arithmetic-logic,* and *output units.* They are arranged as shown in Fig. 4.1.

In this chapter we shall not be concerned with the actual design of the components that make up the computer; we shall, however, in Chap. 8 present some basic notions that are involved in the design of the arithmetic-logic unit. A brief description of the functions of these components would now be appropriate. As you might suspect, the *input* unit is a device by means of which the computer receives the set of instructions it is to subsequently follow and the data, if any, on which the instructions are to act. This information can be fed in via (IBM) cards, from a disk, from a typewriter keyboard, or from a magnetic or paper tape. Once the instructions are stored in the memory unit of the computer, the command is given to start executing the instructions. If all is well and the input has been presented in proper form for the computer to understand, the execution proceeds. The execution can call for the reading in (again by means of an input unit) of data, that is, the numerical values of variables, on which the arithmetic-logic unit is to operate. Following the stored instructions precisely, the control unit passes data and results back and forth between the storage unit and the arithmetic-logic unit where arithmetic calculations are done until all the instructions have been executed. The

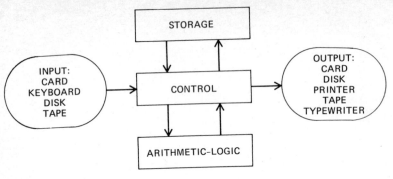

Fig. 4.1

set of instructions should contain the commands necessary to cause the computer to select an output device and to put out the desired results on that device.

As noted above, the computer is an extremely intricate electronic device so engineered and constructed that it will receive, store, and execute your instructions. These instructions must be written and presented to the computer in a manner that it can understand. The only language it can actually understand is called *machine language*. In this language the instructions are a specially coded sequence of (usually) binary digits, and the computer has been constructed to react to these coded instructions in a certain very specific way. Sooner or later if you progress far and deep enough into the design and use of computers, you will have to learn how to write programs in either machine language or a slightly more sophisticated version called *assembly* language. Both of these languages are relatively difficult and tedious to write. To solve problems in applied mathematics and for business and engineering applications, there are many so-called *symbolic* languages that are currently in use by programmers. Among these are Fortran, Cobol, APL, PL1, and myriads of others. All these higher-level languages have the following basic property in common: their use allows the programmer to write his instructions in easy-to-use English and ordinary mathematical words and symbols. This, of course, relieves the programmer from the extremely painstaking work involved when writing in machine language. In order that such symbolic languages can be understood by the computer, it is absolutely necessary that the statements of the program be converted into machine language so that the computer can do its job. So, for each of the languages mentioned above, a special program in machine language has been written by experts which converts the statements of the symbolic language into machine language. This program is called a *compiler*. Suppose a Cobol program, that is, a program written using the Cobol language, has been written and punched in the proper way onto cards. Now, the computer itself can previously have had stored on its disk or in its core memory the Cobol compiler we have just mentioned, whose sole task is to receive the statements of the Cobol program and translate and store the resulting corresponding set of machine language instructions in memory. This chore is usually done completely internally so that one never sees the translation itself. The set of machine language instructions can then be executed.

Each of the languages mentioned above has a compiler associated with it. These compilers are different for different computers since each computer has its own individual characteristics. In most computer centers several of the most used compilers are permanently stored within the computer system so that they can be immediately called up for duty when required. As an additional useful feature, many compilers also contain error-detecting routines that can find certain errors that the programmer may make in the use of his language—these routines usually cause an output listing of the errors they detect.

4.2 INTRODUCTION TO FORTRAN PROGRAMMING

We shall not be able to discuss the details of programming in all these symbolic languages, but we do want to introduce enough details of the basic Fortran IV language so that you can write programs that can actually be run on a computer. In this presentation we cannot go into the important subject of *control cards,* that set of special cards designed to give the computer such information as the language in which the program is written, the number of significant digits the computer should carry in its computations, whether a printed copy of the program should be produced, and many other similar control situations. The format of control cards, which precede and/or follow the program itself, is different for each computer even though the symbolic program itself is written in exactly the same way for different computers. The Fortran programs that we shall learn to write can be executed on any computer that has available to it the proper Fortran compiler. The control cards required for the execution can be supplied by the computer center.

A Fortran program is a sequence of carefully written and ordered Fortran statements. These statements are classed into five categories:

1. Arithmetic statements are used to define the calculations to be performed. We shall study their proper formulation in the next section.
2. Input-output statements are used to transmit information between the computer and its input or output units. We shall present the basic notions of these statements in Sec. 4.4.
3. Control statements are used to govern the sequence of execution of program statements. The most important of these statements will be discussed in Sec. 4.5.

IBM

FORTRAN Coding Form

X28-7327-6 U/M050
Printed in U.S.A.

PROGRAM	NEWTON'S METHOD			PAGE	OF
PROGRAMMER	GEORGE Q. PROGRAMMER	DATE FEB. 6, 1972	PUNCHING INSTRUCTIONS	GRAPHIC	CARD ELECTRO NUMBER*
				PUNCH	

FORTRAN STATEMENT

```
C     THIS PROGRAM SOLVES X**3+4.*X**2-5.*X-11.=0 FOR THE REAL ROOT NEAR
C     X=1 USING NEWTONS METHOD.
      N=0
      X=1
   17 U=2.*X**3+4.*X**2-5.*X-11.
      V=3.*X**2+8.*X-5.
      X=X-U/V
      N=N+1
      IF(N-10) 17,17,10
   10 WRITE(3,36) X
   36 FORMAT(1X,'X','=',E15.6)
      END
```

* A standard card form, IBM electro 888157, is available for punching statements from this form.

Fig. 4.2

4. Specification statements are used to provide information about the data that the program is to process. These statements are covered in conjunction with input-output statements in Sec. 4.4.
5. Subprogram statements are used to define and provide linkage to and from subprograms. We shall not discuss this type of statement in this book.

As you can surmise, there are myriads of details and complications in a full study of the basic Fortran IV language. It is our objective here to present a version that is somewhat simplified but complete in itself. If you care to pursue all the intricacies of the complete language, there are many good texts available (including my book "Fortran IV Programming," McGraw-Hill, New York, 1970).

The set of statements constituting the program are written on specially prepared Fortran coding sheets. Figure 4.2 shows such a sheet with a source program written on it. You note the numbers across the sheet directly under Fortran statement. The numbers correspond to column numbers on a standard 80-column IBM card. Clearly, the Fortran statement must be placed in columns 7 to 72, inclusive. All the alphabetic characters used are capital letters. Once the program is written, one statement per line, on the sheet, it is then the chore of someone, perhaps you, to reproduce each line of the program on IBM cards, punching one line per card, in exactly the same way the statement is written on the coding sheet. The deck of cards thus produced is called the *source deck*. Throughout our discussion of input (of program and data) we shall presume it is done with IBM cards.

Keep in mind that what is being punched on the IBM card is simply a one-to-one reproduction of one line on the Fortran coding sheet. You are thus preparing the source deck which will be read by the card reader. Obviously the computer cannot read a program written on a sheet of paper. All IBM card readers are constructed to read only cards which have been punched "full of holes" by a keypunch. These holes represent a code to the computer; each letter, number, or special character that you "type" on the keypunch keyboard causes a unique hole or set of holes in the corresponding column of the card. Then, when the card is read, the very same letter, number, or special character is interpreted by the card reader. Figure 4.3 shows a punched IBM Fortran card that is in common use.

Let us return to the coding sheet and to the example shown in Fig. 4.2. Notice that the first line contains a *comment*. You can disregard with complete confidence this wholly nonessential feature of programming. However, the heading above the first five columns, STATEMENT NUMBER, is important. These columns may contain a statement number, which is an unsigned integer from 1 to 99999. The statement number can occur anywhere in columns 1

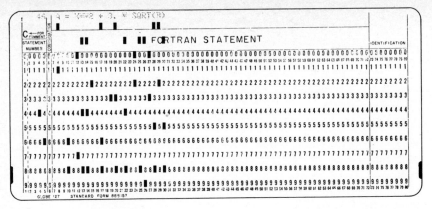

Fig. 4.3 A Punched IBM Fortran Card

through 5. Thus, 37 in columns 3 and 4 would serve to designate the statement number 37. It is usually sufficient to use one or two digits for which it is common practice to use column 5 or columns 4 and 5, respectively. Although any statement may have an assigned statement number, in most cases a statement is numbered only when it is necessary to refer to that statement from some other part of the program. We shall soon see (Sec. 4.5) that a principal use of statement numbers is to provide a reference for control statements. In the example of Fig. 4.2 the control statement GO TO 17 contains the statement number (17) of the next statement to be executed. Not every statement need or even should be numbered. The actual number used as a statement number has absolutely no numerical or arithmetical significance. The numbers are chosen arbitrarily by the programmer. No particular order of numbers is required. Successive statements can be numbered by such a sequence of numbers as 5, 4576, 74, 12, and 3268. The sequence of operations in a program is always dependent upon the order of the statements in the program, not upon the *value* of the statement numbers. One should not, of course, use the same statement number for two different statements.

4.3 FORTRAN ARITHMETIC STATEMENTS

In this section we shall present the proper formulation of arithmetic statements. An arithmetic statement contains constants, variables, arithmetic operation symbols, and an equal sign, and has the *form* of an equation. Let us discuss each of these in order.

A *constant* is a specific number like 8, −1.73, and 7506.45. Constants are of two kinds, *real* and *integer*. A *real* constant is a decimal number which contains a decimal point. Examples are 0.6, −45.6, and 34.56. An *integer* constant is a number that does not contain a decimal point, like 8, −17, and 432. Note that the word "real" as used here does not have the same meaning as it has in mathematics—here a real constant is a number which has a decimal point in it. An integer constant like 13 can be changed into a real constant by simply placing a decimal point after it.

As you recall from our discussion in Chap. 1, the number of digits that these constants can contain is restricted in most computers. This number varies in different versions of Fortran and from computer to computer. For our purposes let us restrict the number of digits according to the following rule: real constants, six figures; integer constants, four figures. This restriction is compatible with current versions of Fortran.

For real numbers like 43800000000., recall the power-of-10 notation 4.38×10^{10}, which we write in Fortran in the form 4.38E+10. Similarly, we write 0.837×10^{-25} as .837E−25. The exponents (like 10 and −25) are also usually restricted in size. Let us again arbitrarily assume that they must be between −30 and +30, inclusive. A general form for real numbers using the E notation is then x.xxxE±xx, where x is a positive digit. Once again, here are some examples of permissible real constants using this notation: −8.765E+00, 72.5689E+23, −.7042E−03. The largest number in absolute value we can write is 1.E+30. Of course, as long as the real number does not require more than six digits to express, we can use the ordinary decimal form for real numbers, like 5.67 and −234.507.

Integer constants, to repeat, are numbers written without a decimal point. Thus, 0, 456, −5069, and 35 are allowable integer constants; 45678 is not permissible since it contains more than four digits.

For both kinds of constants, the plus sign preceding them is optional if the number is positive. Of course, negative numbers are preceded by a minus sign. Incidentally, commas are not permitted in any constant. Thus, one never writes 45,600. in a Fortran statement.

As you have probably expected, just as there are two kinds of constants, there are two kinds of variables called *integer* and *real* variables, so-called because the values that these variables take on are restricted to the corresponding kind of number. It is important to the computer that it be able to differentiate between these two kinds of variables because the mode of arithmetic it will use to calculate expressions involving them is determined by the type of variable the expression contains.

Variables are designated by names as is usual in mathematics. Thus, X, Y, and Z (capitalized) are variable names. In Fortran, though, more complicated multicharacter names are permitted. Thus, DIST, TIME, MASS, M73X, and YY are also variable names. In each of these names the whole set of alphabetic and numeric characters constitute the name. Note that in the *mathematical* expression $xy + 8$, we mean that x and y are separate variables, and xy means to multiply x and y. If we write the Fortran arithmetic expression XY + 8., the computer will assume that only *one* variable is used and that variable has the name XY. Soon we shall see that the proper way to translate the mathematical expression $xy + 8$ into Fortran in order to achieve its usual mathematical meaning is to write X*Y + 8.. Here * stands for the operation of multiplication.

Integer-variable names can contain from one to five alphabetic or numeric characters, but the first character must be alphabetic and must be one of the following letters: I, J, K, L, M, or N. No special characters such as +−(),./* or blanks are allowed in the name of any variable. Thus, M73, IX, J57Y, MM, MASS, and JBM as well as simple one-character names like I,J,K,L,M, and N are allowable integer-variable names. 8N4 is not allowed (its first character is not alphabetic), J**B is not allowed (it contains the special character *), and KIPPERS is not allowed (it contains more than five characters).

A real-variable name may consist of from one to five alphabetic or numeric characters, the first of which must be alphabetic and must *not* be I, J, K, L, M, or N. Thus, A, B, C, X, Y, XX, ZEST, T347A, DSTN, RATE, ALPHA, and R4X5Y are allowable real-variable names. JERK is not permitted since its initial letter is J; RX/BV is not permitted since it contains the special character /; and TY567AS is not permitted since it contains more than five characters.

Note that the computer is using the name merely to identify a certain variable and itself attaches no meaning whatsoever to the name. For example, TIME is just the name of a real variable to the computer, and it does not know that you are using the name TIME in your program to actually represent time. A brief explanation of how the computer treats the variable names it reads in a Fortran statement would be appropriate now. One of the major tasks of the compiler is to note all the constants and variables you have written in your Fortran statements. It causes the constants to be neatly stored. But the variables are designated by names, and they do not have values until the program is executed. For this reason the compiler arranges to have separate and distinct areas of storage reserved for each variable you have named in your program. For example, if you have the variable A in your program, the compiler will reserve an area in storage which it will label A. Then as the values of A are assigned or computed in the course of the execution of the program, they will be stored in the area labeled A. At different times during the execution, the storage area A may have different values of the variable but only one value at any one time. We often say the value is stored at A, when we mean in the storage area labeled A. Suppose that you have written the Fortran expression A + 4.5 in your program. At the time this expression is evaluated, the *current* value of A stored will be brought from storage to the arithmetic unit and used to add to 4.5. What the computer then does with the sum we shall soon learn.

We now turn to the arithmetic operations that are used in Fortran. They are the usual five operations:

Addition, symbolized by +
Subtraction, symbolized by −
Multiplication, symbolized by *
Division, symbolized by /
Exponentiation (raising to a power), symbolized by **

We have already discussed in some detail in Sec. 3.1 the important subjects of hierarchy of operations and the use of parentheses. They both apply to Fortran expressions, so review that material if necessary. Arithmetic expressions are easy to write in Fortran; the only difference from mathematical expressions is that multiplication must be denoted by the asterisk *, division by the slash /, and exponentiation by the double asterisk **. The following examples should be easy to follow:

Fortran expression	Mathematical expression
A + B/C + D	$a + \dfrac{b}{c} + d$
A + B/(C + D)	$a + \dfrac{b}{c + d}$
A**2/3. + B	$\dfrac{a^2}{3} + b$
A + (B/C)**3	$a + \left(\dfrac{b}{c}\right)^3$
(A + B)/(C − D)	$\dfrac{a + b}{c - d}$
A + B*C**X	$a + bc^x$
X**(M − 3)	x^{m-3}

Note that in exponentiation the exponent may be real or integer, constant or variable.

An expression is said to be *real* if all the variables and constants it contains are real. Thus, X*Y + C**3./67. is a real expression. An expression is said to be *integer* if all the variables and constants it contains are integer. Thus, (M/N)**2 + 45/I is an integer expression. An expression that contains both real and integer constants or variables is called a *mixed* expression. The Fortran expression X/N + 56*Y is clearly of this type. All these three types of expressions are allowed in Fortran. Only one circumstance offers any difficulty, the case of the division of one integer constant or variable by another integer constant or variable. An expression like M/N is calculated in what is called *integer mode,* and the result of the calculation is always an integer. Suppose $M = 16$ and $N = 3$. The result of the calculation of M/N will be the *integer* 5. The ordinary arithmetic answer of (approximately) 5.33 is truncated by the removal of the fractional part 0.33. The result is the greatest integer in $16/3$. Another example: 14/8*6 will be calculated as $1 \cdot 6 = 6$, since $14/8 = 1$ using integer arithmetic.

Here is an interesting example which makes explicit use of the property of integer arithmetic concerning division. Suppose we want to determine whether an integer M is divisible by an integer N, that is, N goes evenly into M like 3 goes evenly into 24. We note that if M is not divisible by N, then *ordinary* arithmetic will yield a remainder when M is divided by N. Thus, if M is 17 and N is 6, the quotient would properly be written as $2 + 5/6$; $5/6$ is the remainder. Now if the computer were to evaluate the Fortran expression $17/6$, the result would be 2, and, as we have seen, the remainder would be lost. The fact that a remainder has been lost can be substantiated by noting that the "quotient" 2 multiplied by the divisor 6 does *not* equal the dividend 17. This result also clearly indicates that 17 is not divisible by 6. In general, then, let us call the result of the calculation I by writing $I = M/N$. We then compute I*N. Clearly, M is divisible by N if I*N is M and it is not divisible by N if I*N is not M. For example, suppose $M = 19, N = 4$. Then computing $I = M/N$ yields $I = 4$. Now I*N is $4 \times 4 = 16$. Since 16 is not M, 19 is not divisible by 4. We shall ask later that you write a complete Fortran program concerning divisibility in which you can make use of the material of this paragraph. The division operation of mixed variables will be done using ordinary real arithmetic, and so no difficulty arises. For example, if $X = 32.$ and $N = 5$, the expression X/N will be calculated as 6.4 as expected. Similarly, $48/10. = 4.8$.

One can, of course, use the operation of exponentiation to calculate roots. As you recall, radicals can be expressed using fractional exponents. Thus, $\sqrt{5}. = 5.^{1/2}$, $\sqrt[3]{a + b} = (a + b)^{1./3.}$, and so on. In Fortran one can write $\sqrt{x}$ as X**.5. Be careful, though, about the peculiarity of integer division. For example, one could write $\sqrt[3]{X + Y}$ as (X + Y)**(1./3.) but not as (X + Y)**(1/3). The last expression would be evaluated as 1., regardless of the values of X and Y, since the exponent would be calculated using integer arithmetic and result in the value 0. We shall not be able to go into the special subroutines called the *library functions* that are built into the Fortran system except to mention, for our immediate purposes, the SQRT (square root) function. This library function is used in the following way: SQRT(X). The parentheses are mandatory, and the variable enclosed by the parentheses (the argument of the function) must be either a real constant or a real or mixed expression. It cannot be an integer constant or variable. In any case, the expression SQRT(X) will be evaluated automatically by the computer by taking the square root of whatever the current value of X is. Here are some further examples:

Fortran expression	Mathematical expression
A + SQRT(B + C)	$a + \sqrt{b + c}$
(X**2 + Y**2)**(1./2.)	$\sqrt{x^2 + y^2}$
X**.25/A − (Y + 8.)**.2/7.	$\dfrac{\sqrt[4]{x}}{a} - \dfrac{\sqrt[5]{y + 8}}{7}$

One final note concerning *blanks* before you try the next set of exercises. As you have undoubtedly noticed in the source program of Fig. 4.2 and in many of the example expressions we have written, use has been made of (empty) spaces or blanks so as not to crowd the characters of the expressions together and to provide more clarity and neatness. When the blanks in an expression or statement you write on a Fortran coding sheet are transferred to the corresponding IBM card, the card contains no character at all in that column; no punch appears on the card in the column corresponding to the blank. Blanks in expressions and statements are ignored by the compiler so that, unless a statement is already too long, blank columns can be used to space it. For instance, the statement X=A*B+C/D can be written X = A * B + C / D, and the statement GO TO(12,13,14,15),M can be written GO TO (12, 13, 14, 15), M. One should not insert blanks in the middle of a number; one should not write X = 8 3.0 when one means, algebraically, $x = 83$. And, of course, one cannot have blanks in the name of a variable since the computer then considers the blank to be a special character.

EXERCISES

4.1 Identify each of the following as a real or an integer constant. If the constant is not acceptable (according to the criteria given in the text), state why.

(a) −457
(b) 0
(c) 0.0034
(d) 328.8934
(e) 67.E+03
(f) −677308
(g) +87
(h) 3.E−89

4.2 Write each of the following real numbers in exponent form with the decimal point just preceding the first significant nonzero digit.

(a) .0678E+04
(b) 876.1

(c) 9.35E−06

(d) 0.0000000067845

(e) 5678.4E+12

(f) .000347E−08

4.3 In the following list identify the acceptable variable names as real or integer. If the name is not acceptable in Fortran, state why.

(a) V

(b) LAMBDA

(c) SUM7

(d) *AB

(e) N45X

(f) TIME

(g) XNNN

(h) LAST

(i) FIRST

(j) SECOND

(k) 57Y

(l) XXXXX

(m) X/UZ

(n) P5678

(o) 45.6

4.4 Write a Fortran expression for each of the following mathematical expressions. Use the same variable names as those used in the expression.

(a) $a - \dfrac{b+c}{8}$

(b) $\left(\dfrac{p-q}{p+q}\right)^3 - 7$

(c) $a + (b-c)^{4/5}$

(d) $\sqrt{u^3 + v^4}$

(e) $a + \dfrac{b}{c} + \dfrac{d}{e}$

(f) $a - \dfrac{b}{4-x}$

(g) $\dfrac{1}{5}\left(\dfrac{x+3}{n} + k\right)^j$

(h) $\dfrac{5 + (24/t)}{\sqrt{s}}$

(i) $\dfrac{3470}{d^2} + \dfrac{7856}{17r}$

(j) $n\left(d^2 + \dfrac{c}{m}\right)$

(k) $1 + x + x^2 + x^3 + x^4$

4.5 Shown below are pairs of expressions, each of which consists of a mathematical expression and what purports to be the corresponding Fortran expression. Each contains at least one error. Point out the errors and correct the Fortran expressions.

(a) $(r/s)^{k+7}$ (R/S)**K+7

(b) $(5 + p)^3$ 5. + P **3

(c) $\dfrac{x + 7}{y - 6}$ X + 7/Y − 6

(d) $\left(\dfrac{d + 5}{d - 3}\right)^4$ (D + 5.)/(D − 3.)**4

(e) $a(x + y)$ A*X + Y

(f) $\dfrac{a}{c} + \dfrac{c \cdot d}{a}$ A/C + CD/A

(g) $\dfrac{6}{xyz}$ 6./X*Y*Z

(h) $\dfrac{45{,}678 + p}{5{,}557 - q}$ (45,678. + P)/5,557. − Q

4.6 Determine the mode of the following expressions (real, integer, or mixed):

(a) I + 2/NXX

(b) X*Y + 34.5/P78

(c) X + 2 + 3.*Q

(d) I**3 + (N + 5)/X

(e) 456 − NNNN + NP/6

(f) X**3. − X**2. + 5.*BIG

We have finally arrived at the main objective of the section — the writing of Fortran arithmetic *statements*. Every Fortran arithmetic statement has the following form: variable = expression. The variable may be real or integer (and, naturally, is not a constant or an expression), and the expression may be a variable, a constant, or any properly formed Fortran arithmetic expression. Thus, V = X*Y**4 and N = X + M − 7.8 are correctly formed statements, but 4 = X and C + D = P**3 are not. Note that C + D cannot be the name of a variable since it contains the special characters blank and +.

This is the way an arithmetic statement is executed. The expression on the righthand side of the equal sign is evaluated using the mode of the expression itself—integer for an integer expression, real if the expression is real or mixed. This evaluation takes place regardless of the mode of the variable on the lefthand side of the equal sign. The result is changed in mode, if necessary, to agree in mode with the variable on the left, and it is stored at that variable's storage location. Thus, the execution of an arithmetic statement changes the value of only one variable, the one on the left of the equal sign. All the variables in the righthand expression keep the *same* values they had. Thus, for $N = X + Y - 5.0$, we have the following: Add x to y, subtract 5.0, and store the result at n. Since N is an integer variable, its values are integers, and an integer must always be stored at N. Once the real expression $X + Y - 5.0$ has been calculated in real mode, the decimal part of the number (together with the decimal point) is removed, and the resulting integer is stored at N. Suppose x were 8.47 and y were 6.63; the real result would have been $8.47 + 6.63 - 5.0 = 10.10$, and 10 would be stored at N.

Similarly, when the statement $X = N + 8$ is executed, the integer result of adding N to 8 is made real by the proper insertion of a decimal point, and the corresponding real number will be stored at X.

A very simple way of initially assigning a value to a variable is to write a statement like $X = 3.$. When this statement is executed, the value 3. will be stored at X. The value stored at X need not always remain 3. If, later in the program, an arithmetic statement has X as its left-side variable, the new value of X thus calculated will simply replace the old value of X. The old value of X is now erased and is lost. Only one value of X may be stored at any one time in the storage locations reserved for the values of the variable X. Consider the following sequence of arithmetic statements:

$A = 4.0$
$M = 3$
$A = A + 2*M$

When these statements are executed, the value of A will be initially set at 4.0, the value of M initially set at 3. The righthand side of the third statement will now be executed: the current value of A (namely, 4.0) will be added to $2*M$ (which is 6) with the result 10.0. Then, since A is the variable on the left side of the statement, the new value of 10.0 will be stored at A, erasing the old value of 4.0. stored there by the first statement.

This last sequence of statements has made us realize that Fortran arithmetic statements are not equations in the algebraic sense. Rather, they contain what is to be calculated and where the result of the calculations is to be stored. The algebraic equation $n = n + 1$ is not true for any n. But the Fortran statement $N = N + 1$ is not only perfectly valid but turns out to be rather useful. Let us analyze this statement. Suppose that a value of N is stored previous to the execution of the statement. The expression N + 1 is evaluated, that is, 1 is added to N, and the result is stored at N. Clearly, the effect is to advance the number stored at N by 1. The original value of N is lost. The number stored at N can be thought of as a counter in many applications. In a similar way, the statement SUM = SUM + X advances the value of SUM by X.

Here are some examples of properly formed Fortran arithmetic statements of the mathematical statements they represent.

$a = 5(r^3 + 4rh)$	A = 5.*(R**3 + 4.*R*H)
$n = d^g + g^d$	N = D**G + G**D
$t = \sqrt[15]{x + 7}$	T = (X + 7.)**(1./15.)
$u = 16\alpha^2 + \sqrt{\beta}$	U = 16.*ALPHA**2 + SQRT(BETA)
$w = \dfrac{68}{4/x + 7/y}$	W = 68./(4./X + 7./Y)
$d = \sqrt{b^2 - 4ac}$	D = (B**2 − 4.*A*C)**.5

It should be clear to you now that all the numbers the computer uses to calculate with are *rational* numbers. If you ask the computer to calculate $\sqrt{31}$ by writing 31.**.5, the six-significant-figure rational approximation 5.56776 will be the best effort of the computer. As you recall from the discussion in Chap. 1, even some rational numbers like 1./3. require infinitely many digits to be represented exactly whether they are stored as decimal numbers or binary numbers, and since only a finite number of these digits can be used to represent the number, calculated results will contain errors due to truncation. Exactly the same kind of error due to truncation occurs in any computing device because all computers have only a finite number of storage places to store the digits of a number.

We have learned in this section how to properly write Fortran arithmetic statements that will cause the calculations of the operations of addition, subtraction, multiplication, division, and exponentiation to be performed when a program containing them is executed. We have also seen the use of arithmetic statements to assign values to variables. The prime way, however, of assigning values to variables is to read in numerical data during the execution of the program. In the following section we show how one prepares data cards, and we present the input Fortran statements that will be required in the program so that data cards can be read in at the appropriate time. We shall then discuss the outputting of results. After the introduction of just a few more essential Fortran statements, we shall finally employ all we have learned to write complete Fortran programs.

4.7 State the value of X or I stored as the result of each of the following arithmetic statements.

(a) $X = 2*6 + 3$
(b) $I = 2*40/6$
(c) $I = 2/40*6$
(d) $X = 2*40/6 + 3$
(e) $X = 50/3$
(f) $X = 2. + 37/5$
(g) $X = 15/6 + 17/2$
(h) $I = 3.7 + 8.7 - 4/3$

4.8 Write arithmetic statements to compute the following formulas. Use Fortran coding sheets, if possible, to write the statements on. Use the letters in the formulas for variable names.

(a) $y = \sqrt{2x + 7}$

(b) $\theta = \dfrac{5x^3}{(a^2 + b^2)^{4/3}}$

(c) $v = 3.14159r^2$

(d) $v = \dfrac{1}{3}(3.14159)r^2h$

(e) $q = \dfrac{h}{6}(v^7 + 3w^6)$

(f) $h = \dfrac{q + \sqrt{6}}{q - \sqrt{5}}$

(g) $r = \sqrt{2 + \sqrt{2 + s}}$

(h) $t = \dfrac{a^b}{6} - \dfrac{c^d}{10}$

(i) $\alpha = \dfrac{4}{\beta}(\beta^2 - 4)^{3/2}$

(j) $z = \left(\dfrac{4 - x}{5 + y} - \dfrac{4 + x}{5 - y}\right)^3$

4.4 FORTRAN INPUT–OUTPUT STATEMENTS. FORTRAN PROGRAMS

We have seen that variables can be assigned values by using arithmetic statements in the program itself. Every time such a program is run, the same values will be assigned to the same variables, and the results of the program will inevitably be identical. Most programs are designed so that the values of at least some of the variables they contain are different each time the program is executed. Instead of writing a sequence of arithmetic statements like

A = 4.7
X = 8.32
C = A**2 + X**3

one could write the *input* statement READ (2, 5) A, X followed by the statement C = A**2 + X**3. A card containing the values of A and X would be read by the card reader, and these values used to calculate C. Then, if this sequence of statements is executed again, it is perfectly possible that the new card will have different values of A and X punched on it. The card being read which contains the values of A and X is called a *data card*. Notice that the data cards are not part of the source program; they are read and entered by the computer after the program itself has been written, punched, compiled, and stored and is in the process of being executed.

In this section we shall learn how to prepare data cards, how to properly write input READ statements and their associated FORMAT statements. We shall also have to learn to write proper *output* statements in order to discover the results of the calculations of the program; we shall describe various output WRITE statements and their associated FORMAT statements. In this book we shall be concerned only with the input and output of *numerical* data. A full description of the Fortran IV language would include a presentation of the input and output of alphanumeric data, that is, data that could contain alphabetic and special characters as well as numeric characters. In keeping with the fundamental character of our entire presentation, we shall restrict even our discussion of input and output to basic essentials. We shall assume throughout that the input device being used is the card reader and that the output device is the printer.

The input statement has the form

READ (J, K) list

where J is a logical unit number (an integer constant or variable) indicating which input device is being used, K is an integer constant which is the statement number of the associated FORMAT statement (of which we shall have much more to say presently), and *list* is the list of the variables whose values are to be inputted. The comma between J and K is mandatory, and the different variable names in the list must be separated by commas, but no comma is placed after the last variable in the list. The READ statement thus indicates that input is to be instituted, tells what input device is to be used and what the names of the variables whose values are to be read are, and refers to a FORMAT statement which, as we shall see, specifies the type and length of the numbers.

Let us consider J, the code which identifies the input device to be used. Different computer systems use varying input devices and varying codes to identify them. For our purposes we shall use 2 for J and assume that this value for J identifies the card reader as the input device. Thus, the statement READ (2, 56) N, X, Y, P indicates that the values of the variables N, X, Y, and P are to be entered by means of a punched card read by the card reader. The 56 refers to the FORMAT statement whose number is 56 and which contains the necessary information about how these values are punched on the card.

Let us now turn to the subject of the data card itself. Recall that a standard IBM card contains 80 columns in which the values of the variables are to be punched. Since the card is always read from left to right, the values of the variables must be in the same order as the list of names so that the proper value will be assigned to each variable. Except for this order, the values may appear anywhere on the card. The FORMAT statement will tell the type of numbers they are and where they are on the card. Here is an example of a READ statement and its corresponding FORMAT statement.

READ (2, 47) P, M, Z, W, J, R, I
47   FORMAT (F6.2, I4, E6.1, F10.4, I2, F12.4, I3)

Notice that within the parentheses of the FORMAT statement there are seven *specifications*, one for each variable of the list. The commas which separate the specifications are mandatory punctuation. Let us first explain the special code letters F, E, and I which occur in this statement. F identifies a real number written in ordinary decimal form; E identifies a real number written with the exponential notation; I identifies an integer number. As you look over the FORMAT statement, notice that each real variable of the list has either an E or F specification and that each integer variable of the list has an I specification. This correspondence is required. The integral part of each number that follows these code letters indicates the number of columns of the card the corresponding punched number may occupy. Thus, the 6 in F6.2 shows that the value of the variable P is contained in 6 columns of the card. Four columns are used for M, and so on across the card. Thus, the sequence 6, 4, 6, 10, 2, 12, 3 of these integral parts divides the card into consecutive sets of columns each containing the value of the corresponding variable. The card could look like the card of Fig. 4.4. The values of the variables appear on the card in this way: P in columns 1 to 6, M in columns 7 to 10, Z in columns 11 to 16, W

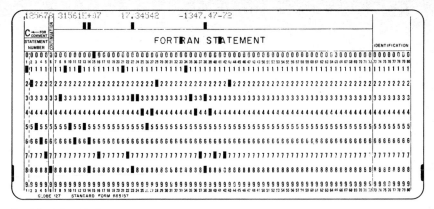

Fig. 4.4 A Data Card

in columns 17 to 26, J in columns 27 and 28, R in columns 29 to 40, and I in columns 41 to 43.

Note that the value of M as punched in columns 7 to 10 must be *right-justified,* that is, the last digit of M must appear in column 10.   Suppose M were 89; then

$$\underset{\underset{\text{cc7}}{\uparrow}}{} \, \underset{}{} \, \underset{}{\underline{8}} \, \underset{\underset{\text{cc10}}{\uparrow}}{\underline{9}}$$

is correct.   If the card were punched

$$\underset{\underset{\text{cc7}}{\uparrow}}{} \, \underset{}{\underline{8}} \, \underset{\underset{\text{cc10}}{\uparrow}}{\underline{9}} \, \underset{}{}$$

with a blank (no punch at all) in column 10, the computer would automatically assume a *zero* in column 10 and would read the value of M as 890.   To avoid such errors, it is best for all the numbers punched on the card to be right-justified so that the last digit of the number appears in the last column of the field of columns assigned for that number.   Thus, if Z were 8.32, we would have

$$\underset{\underset{\text{cc11}}{\uparrow}}{} \, \underset{}{} \, \underset{}{\underline{8}} \, \underset{}{\underline{.}} \, \underset{}{\underline{3}} \, \underset{\underset{\text{cc16}}{\uparrow}}{\underline{2}}$$

In the case of F and E specifications, the digit following the decimal point denotes how many decimal places the number is to contain.   This part of the specification is ignored if the decimal point is *physically* punched on the card. It can be used, however, to place the decimal point if one is not punched. In columns 1 to 6, for example, 125678 is punched. The specification is F6.2, so when the card is read the value of 1256.78 will be assigned to P. Similarly, in columns 11 to 16, 61E+07 is punched. Since the specification is E6.1, the value 6.1E+07 will be assigned to the variable Z. In the case of W, the specification is F10.4. The 4 in F10.4 will be ignored since the decimal point is already punched in 17.345 and 17.345 will be assigned as the value of W. The following values will thus be assigned to the variables of the READ statement's list when the card is read: P = 1256.78, M = 315, Z = 6.1E+07, W = 17.345, J = 42, R = −1347.47, and I = −72.

With reference to the preparation of data cards, two situations often confront one. In the first case, one may be asked to write a general-purpose program to solve a certain problem that will require the reading of data the values of which one would not know in advance when writing the program. In this circumstance one can write the FORMAT statements in an appropriate manner which will surely accommodate the expected data, and whoever uses the program must of course prepare his data cards to conform with the given specifications. An example of this kind of program could be one to solve the general quadratic equation $ax^2 + bx + c = 0$ in which the data read in would be the values of a, b, and c. In the second case, one may be writing a very specific program with data on hand that one knows. One may then write the FORMAT statements in any manner that is appropriate and then follow those specifications as one prepares the data cards for immediate use.

There are many more facets and techniques of input that we shall not consider here. Again we assume that if you desire to delve more deeply into this subject, you consult a complete text on Fortran programming.

Now we turn to the problem of the output of the results of calculations or other desired numerical items. Recall that we have selected the printer as the output device. Again we shall present just the amount of fundamental output material to enable us to achieve output in the simplest manner. The Fortran output statement is the WRITE statement. It and its associated FORMAT statement have the following form:

```
        WRITE (M, N) list
    N   FORMAT (    )
```

Here the word WRITE indicates that data are to be outputted, M is the logical code number of the output device (we shall use 3 to indicate the printer), N is the statement number of the associated FORMAT statement, and *list* is the list of variables whose values are to be outputted. The FORMAT statement prescribes the layout of the printed line. Here is an example.

```
        WRITE  (3,  18)  X,  Y,  N
  18    FORMAT  (1X,  F12.5,  30X,  E10.3,  I5)
```

Here is a description of the functions of the various items that appear within the parentheses of the FORMAT statement. We immediately recognize the code letters F, E, and I that are clearly associated with the variables X, Y, and N, respectively. In output, however, the digits that follow these letters do not have the same significance as for input since they are evidently trying to prescribe just how the printed line will look. In input the specifications in the FORMAT statement are completely determined by the manner in which the data card is punched. In output the numbers to be printed are in storage deep in the computer, and their values and magnitudes may be completely unknown to you.

If you output a real number under the specification Fw.d, the number will be outputted with d places to the right of the decimal point and will be in the form of an ordinary decimal number. The w in Fw.d is the number of printing positions that will be used to print the number. When using the F specification to output a real number, one must be careful to allow enough printing positions to accommodate the number. Suppose the number 79.56 is to be outputted. One could use the specification F6.2 because exactly six printing positions will be required. The number (including the printing of its decimal point) will require five positions, and one position will be used for the sign of the number. The plus sign is not printed if the number is positive, but a printing position must be reserved for the sign. If the number is negative, a minus will be printed. Any specification larger than F6.2, such as F9.2, will clearly accommodate the number, and the printing under this specification will cause three blank spaces before the printing of the number itself. If you write F7.3, then three places to the right of the decimal point will be printed, and the entire number will be printed as 79.560. Note that w must now be at least 7; the zero on the end now requires an additional printing space.

Many Fortran compilers will cause the printing of a set of asterisks instead of the value of the number if the specification is not the proper size to accommodate the number to be outputted. Suppose you desire to output 567.893 and you provide the specification F5.3. Clearly, the number cannot be outputted in such a few numbers of printing positions. The computer could cause the output to be *****, which would be disconcerting, to say the least. As you have no doubt surmised by now, it is not always evident in advance just what the size of the number to be outputted is going to be. This asterisk type error is then liable to occur rather often. There is, however, a way out which will never fail: using the E specification, which we now discuss.

The Ew.d specification as usual calls for w printing positions to be used and for d decimal places to the right of the decimal point. In addition, it calls for the number to be printed in exponential notation. The E specification always causes the normalizing of numbers on output; that is, whatever the number is, its decimal point is shifted to the immediate left of the first nonzero significant digit of the number, and the power of 10 following the number is adjusted to compensate for this shift. Suppose we have the number 874.56 stored. When this number is normalized, it becomes .87456E+03. Under the specification E12.5, this number would be printed as 0.87456E+03. We note that 12 printing positions are used. One is used for the sign (this will be printed if the number is negative), two are used for 0. (this 0 is always printed), five are used for the digits of the number, and four printing positions are used to print E+03. Now you recall that the maximum number of digits a real number can contain is six. From the counting that we did in the last example, it should be clear to you that in order to output a real number of six digits the specification E13.6 is entirely sufficient. As noted before, a specification like E17.6 will merely cause four spaces to appear before the number itself. Thus, we shall follow the rule that unless we definitely know just about what the real number is that we want to output, we shall use (at least) the specification E13.6. If we know the number is about 13.45, we can of course use an F specification like F7.2.

This last mention of the occurrence of spaces brings up an important point. We would prefer that the various values of variables that appear in a printed line have spaces between them. This horizontal spacing of numbers across the line can be achieved by increasing the w of the specifications as we have indicated above. A simpler way of spacing, however, is to use the specification wX. Here X is a code letter and w is the number of spaces desired. Note the previous example:

```
    WRITE (3, 18) X, Y, N
18  FORMAT (1X, F12.5, 30X, E10.3, I5)
```

Evidently, one space will occur in the printout just before the value of X is printed, 30 spaces will occur between the printing of the values of X and of Y. The printed line has a maximum number of printing positions, usually 120. The FORMAT statement just given will cause the values of X, Y, and N to be printed on one line.

The material presented so far concerning input and output is sufficient for our immediate purposes. More will be revealed in the chapter on arrays and matrices.

We introduce just one more Fortran statement before we start writing some complete programs. Some method must be found to signal the compiler that the end of the program has been reached, that it can stop reading Fortran statements, and prepare to do its compiling and translating job. This is done using the END statement: we simply write END. Every Fortran program *must* have this statement as its last statement. Now for some sample Fortran programs.

Example 4.1. Write a complete Fortran program to assign the value 6.789 to the variable X, 7 to the variable M, -34.66 to the variable P, compute and output the value of the variable Y if $y = \sqrt{x^3 + p} + (x - p)^m$.

Analysis: The plan of this program is direct and simple. We shall use arithmetic statements to assign the values of the variables, write an arithmetic statement to compute y, and follow it by an appropriate output statement to print the value of y, as follows.

```
      X = 6.789
      M = 7
      P = -34.66
      Y = SQRT(X**3 + P) + (X - P)**M
      WRITE (3, 45) Y
   45 FORMAT (3X, E13.6)
      END
```

It is characteristic of computer programs that the various statements of the program are executed in order, that is, sequentially, unless somehow the order is altered by some control statement. The program above is clearly sequential. The statement which calculates the value of y cannot be executed unless the computer had been previously informed as to the values of x, p, and m. This has luckily been done in the first three statements. Note that the FORMAT statement contains the useful specification E13.6.

Example 4.2. Write a complete Fortran program to read a card containing the values of the variables A, B, C, and D, compute $x = a/(b + c)$ and $y = d^2 + c^3 + 4x$, and print the values of a, b, c, d, x, and y on one line.

Analysis: Since the FORMAT of the card must be known, let us assume that someone has informed us that the value of A is punched in the first 10 columns, the value of B in the next 12 columns, the value of D in the next 10 columns, and the value of C in the next 8 columns, and that each is in decimal form with the decimal point punched. The program appears as follows.

```
      READ  (2, 35)  A,  B,  D,  C
 35   FORMAT  (F10.0,  F12.3,  F10.1,  F8.0)
      X = A/(B + C)
      Y = D**2 + C**3 + 4.*X
      WRITE  (3, 568)  A,  B,  C,  D,  X,  Y
568   FORMAT  (E15.6,E15.6,E15.6,E15.6,E15.6,E15.6)
      END
```

The input format statement should be clear. The computer then knows
the values of a, b, c, and d when it comes to the third statement. It also knows
the value of x when it is asked to calculate the value of y. We have used E15.6
as the output specification for all the variables, since it allows for horizontal
spacing without the use of the X specification. It is possible, though not
necessary, to simplify this FORMAT statement by writing it this way: 568
FORMAT (6E15.6). The 6 preceding E15.6 merely indicates that the specification E15.6 is to be repeated six times. Notice that the entire program has
been written without the knowledge of the actual values of A, B, C, and D.
If desired, then, this program can be run over and over again with each run
using a different set of values for these variables, that is, with different data
cards.

Example 4.3. Write a complete Fortran program to solve the following set of
simultaneous linear equations:

$$\begin{cases} ax + by = c \\ dx + ey = f \end{cases}$$

The values of a, b, c, d, e, and f are to be read on *two* cards, the first card containing the values of a, b, and c punched in format 3F7.2, the second card containing the values of d, e, and f punched in format 3F10.0. Print the values of
x and y on *consecutive* lines.

Analysis: We leave it to you to show that the value of x is $(ce - bf)/(ae - bd)$
and the value of y is given by $(af - cd)/(ae - bd)$. We shall assume that
$ae - bd \neq 0$. We first present the program and then give a discussion of
some of its interesting facets.

```
      READ  (2, 1)  A,  B,  C
  1   FORMAT  (3F7.2)
      READ  (2, 2)  D,  E,  F
  2   FORMAT  (3F10.0)
      DENOM = A*E - B*D
```

```
      X = (C*E − B*F)/DENOM
      Y = (A*F − C*D)/DENOM
      WRITE (3, 3) X
  3   FORMAT (1X, E13.6)
      WRITE (3, 3) Y
      END
```

With the simplified version we are presenting, the *two* READ statements are necessary since we are to read the data from two *separate* cards.

```
      READ (2, 1) A, B, C, D, E, F
  1   FORMAT (3F7.2/3F10.0)
```

The above set of two statements can replace the first four original statements. The slash (/) in the input FORMAT statement is a code to the computer that the reading of the card is to end at that point and another card is to be read for more data. In this context / does not mean division. The use of this additional coding merely simplifies the writing of the input statements for the program and we can get along without it.

Note that the expression $ae − bd$ occurs in the denominators of both x and y. It is good practice to compute this quantity separately *once* (here it is called DENOM), rather than to compute it twice, once when x is calculated and another time when y is calculated.

It is not clear why the two WRITE statements will cause X and Y to be printed on consecutive lines. The 1X which appears in the FORMAT statement numbered 3 is a special code to the printer that a new line of printing is to be initiated. We shall continue to use this printer signal 1X as the first item of the output FORMAT statements we write in order to ensure that printing will always occur on new lines. The 1X code we use here may not be the code for every computer or compiler.

Another way to write the set of two WRITE statements is as follows:

```
      WRITE (3, 3) X, Y
  3   FORMAT (1X, E13.6)
```

This set of statements will cause the printing of X and Y on separate lines. First, X will be printed (on a new line because of the specification 1X). The specifications are now all used up, and there appears to be no specification for the variable Y. However, when this occurs, the FORMAT statement will be again entered at the left parenthesis and the specification will be reread and used for the variable Y. Evidently, the 1X will cause the printer to space vertically so that the value of Y will be printed on a new line. If there were more variables in the list, this process would be repeated for each of them. Perhaps you are wondering whether such rereading of specifications could be used in input too. It can. For example, consider

```
    READ  (2, 7)  X, Y, Z, P
  7 FORMAT  (F10.0)
```

In this case, each time the FORMAT statement is reentered, a *new card must be read*. Thus, this set of statements will cause the reading of four cards; each of the cards should contain the value of the corresponding variable in the first 10 columns of the card, and each card is read under the specification F10.0.

In the program above the output FORMAT statement numbered 3 is referred to by both the WRITE statements. This is perfectly legal. In addition, FORMAT statements need not necessarily follow the READ or WRITE statement that refers to them. All the FORMAT statements of the program may be placed in a set at the beginning of the program. We prefer the natural order as used here.

Let us present the rewritten program above using the simplifications just remarked on.

```
    READ  (2, 1)  A, B, C, D, E, F
  1 FORMAT  (3F7.2/3F10.0)
    DENOM = A*E − B*D
    X = (C*E − B*F)/DENOM
    Y = (A*F − C*D)/DENOM
    WRITE  (3, 3)  X, Y
  3 FORMAT  (1X, E13.6)
    END
```

Again, note that this program can be used to solve many sets of simultaneous linear equations. The first two cards read can be different on each run of the program. At this point, however, the only way the program can be rerun is to start again from the reading in of the source deck and proceeding through compiling and execution each time. In the following section we shall present *control* statements that can be inserted in this program and which will cause the immediate rerunning, in essence, of the program. As you recall, control statements can alter the usual order of execution of statements.

EXERCISES

4.9 Figure 4.5 shows a punched data card which is read using the statement READ (2, 4) A, B, N, X, Y, M. What are the values assigned to these variables if the corresponding FORMAT

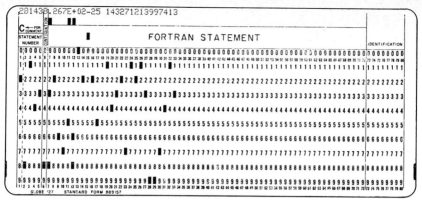

Fig. 4.5 A Data Card Called for in Exercise 4.9

statement is:

(a) 4 FORMAT (F5.2, E9.1, I3, F7.3, F6.5, I2)
(b) 4 FORMAT (F7.5, E7.1, I4, F5.2, F6.2, I3)
(c) 4 FORMAT (F4.0, E10.1, I3, F6.2, F6.4, I3)
(d) 4 FORMAT (F6.1, E8.5, I2, F7.2, F5.4, I4)

4.10 Given the READ statement READ (2, 5) A, B, C, D, E, F, G, H, under each of the following FORMAT statements tell how many cards are to be read and which variables are to be on each card:

(a) 5 FORMAT (8F10.0)
(b) 5 FORMAT (3E12.6, 3F10.0, 2F7.5)
(c) 5 FORMAT (F12.0)
(d) 5 FORMAT (2E13.6)
(e) 5 FORMAT (4F10.0/4E15.5)
(f) 5 FORMAT (2F10.0/4E13.6/F7.2, F5.1)

4.11 Write a complete Fortran program to assign the value 4.72 to the variable X, -12.73 to the variable Y, 7 to M, and -5 to N, compute $z = x^2 + y^{m+n}$ and $q = \sqrt{z} + (x^m + y^n)/17$, print z and q on one line.

4.12 Write a complete Fortran program to read a card containing the values of the variables A, B, C, D, and E. The card is punched 5F12.0. Compute $x = \sqrt{a^2 + b^2 + c^2 + d^2}$, $y = \sqrt{x^2 + e^3}$, and $z = (4x + 5y)/2$. Print A and B on one line, C, D, and E on the second line, and X, Y, and Z on the third line.

INTRODUCTION TO SYMBOLIC PROGRAMMING LANGUAGES 115

4.13 Write a complete Fortran program to read two data cards. The first card contains the values of X, P, Q, and N. X, P, and Q are punched as decimal numbers with the decimal point punched. X appears in columns 1 to 10, P in columns 11 to 15, and Q appears in columns 16 to 25. N is punched in columns 26 to 29. The second card contains the value of I punched I3. Compute $a = (x - 4p + 7q)/11$, $b = \sqrt[3]{q^n + p^i}$, and $c = a^2/5 + 56p - \sqrt{ab}$. Print A, B, and C on one line.

4.14 Write a complete Fortran program to read three cards with the values of A, B, and C, each punched F9.0. A, B, and C are all positive numbers and represent the lengths of the sides of a triangle. Compute the semiperimeter and area of the triangle using the formulas $S = (a + b + c)/2$ and $AREA = \sqrt{s(s-a)(s-b)(s-c)}$. Print A, B, and C on one line. Print S and AREA on the next line.

4.5 FORTRAN CONTROL STATEMENTS

Control statements in Fortran are those that will enable you to control the course of the program. As you recall, statements are normally executed sequentially; after one statement is executed, the statement immediately following it is executed. It is often more useful to alter the sequence of the statements of the program, as we shall soon see. In this section we shall present the two basic control statements (unconditional) GO TO and IF. There are additional control statements in the expanded version of Fortran.

The GO TO statement provides a means of transferring control immediately to some other statement of the program. This is often immediate *branching*. It is written in the form

GO TO n

where n is the statement number of the statement to which the control is to be immediately transferred. This means that when the statement GO TO n is executed, the next statement executed is the one numbered n. This will happen whether the statement numbered n is before or after the GO TO statement itself.

An obvious use of the GO TO statement is to return to the beginning of a program so that the complete program can be executed again. The following example shows an important process in programming called *looping*. Consider

```
75   READ(2, 235) X, Y
235  FORMAT  (2F10.0)
     Z = X**2 − 4.*X/Y
     WRITE  (3, 17) Z
17   FORMAT  (1X, E13.6)
     GO TO 75
     END
```

When executed, this program will first cause a card with the values of X and Y to be read, the value of Z calculated and printed.   Then control is transferred by means of the statement GO TO 75 to the statement numbered 75.   This means that the READ statement will be executed next, a new card will be read, a new value of Z calculated and printed.   Again control is transferred to the READ statement and the process will be repeated.   Since now a *loop* has been created, the process will be repeated over and over again as long as there are cards supplied to be read.   Note that the presence of the END statement has no effect on the looping; it does not cause the program to end. We shall return to this program once again to see how we can limit the number of cards that will be read, if that is our plan.

In any case, the GO TO *n* statement alters the sequence of statement execution unconditionally; the transfer of control is immediate and does not depend at all on any values computed in the program.   The *n* in the GO TO *n* statement must be the statement number of *some* executable statement in the program; if no statement is numbered *n*, the GO TO is rather pointless, and its presence in the program reveals a certain degree of poor planning on the part of the programmer.   Note that it is not permissible to transfer control to a nonexecutable statement like FORMAT or END.

Before we present some more examples of the use of GO TO and introduce the IF statement, we consider the concept of the flowchart.   A *flowchart* is a graphical presentation of the order of execution of the statements of a program.   It should be carefully drawn for each program in order to plan the logical flow of the program before writing out the statements of the program in detail.   Figure 4.6 shows the graphical symbols we shall use to create the flowchart.   To illustrate the use of the flowchart, we diagram the last illustrative program in Fig. 4.7.   Note the arrow indicating the return to read a new card.   Note also that the complete statements of the program itself are not given in the diagram.   The intricate details of the actual written program are to be supplied by the programmer after he has planned the desired flow of execution with a flowchart.   You realize, of course, that we wrote the program without the aid of a flowchart since it was so simple in its flow.   But in even moderately complex problems the various interrelationships within the

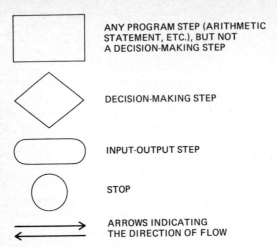

ANY PROGRAM STEP (ARITHMETIC STATEMENT, ETC.), BUT NOT A DECISION-MAKING STEP

DECISION-MAKING STEP

INPUT-OUTPUT STEP

STOP

ARROWS INDICATING THE DIRECTION OF FLOW

Fig. 4.6 Flowchart Symbols

program often become very difficult to keep clear without the use of some visual scheme.   In addition, appending the flowchart to a complex program is a valuable aid in communicating to others the actual operation of the programming.   The flowchart is thus a part of the *documentation* of the program.

A CONTROL statement that involves decision making is the IF statement.   It has the form

IF (E) n_1, n_2, n_3

Here E stands for any expression, and n_1, n_2, and n_3 are statement numbers (and are thus unsigned integer constants).   This is the way the statement works.

Fig. 4.7 Flowchart

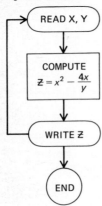

READ X, Y

COMPUTE
$Z = x^2 - \dfrac{4x}{y}$

WRITE Z

END

When the IF statement is executed, the current numerical value of the expression is calculated. Then if this value is less than zero (that is, negative), control is transferred immediately to the statement numbered n_1; if the value of the expression is zero, control is transferred to the statement numbered n_2; and if the value of the expression is greater than zero (that is, positive), control is immediately transferred to the statement numbered n_3. Thus, the IF statement causes a decision to be made on where to transfer to, based on the current value of E. As we shall see, when you write the IF statement in a program, you should already have planned to cause the program to branch to various places in the program for various purposes. The statements numbered n_1, n_2, and n_3 must appear somewhere in the program, and they can be before or after the IF statement itself. However, the numbers n_1, n_2, and n_3 need not necessarily be distinct. The statement

IF (Y) 13, 13, 14

is perfectly permissible. The three statement numbers must be listed, separated by commas as shown. Some examples follow.

Example. Write an IF statement that will transfer control to statement number 73 if $B^2 - 4AC < 0$, to statement number 86 if $B^2 = 4AC = 0$, and to statement number 412 if $B^2 - 4AC > 0$.

Solution: IF(B**2 − 4.*A*C) 73, 86, 412

Example. Write an IF statement that will transfer control to statement number 12 if $X > Y$ and to statement 15 if $X \leqslant Y$. Note that $X > Y$ means that $X - Y > 0$ and $X \leqslant Y$ means $X - Y \leqslant 0$.

Solution: IF (X − Y) 15, 15, 12

Example. Write an IF statement that will transfer control to statement number 7 if $N \geqslant 9$ and to statement number 18 if $N < 9$.

Solution: IF (N − 9) 18, 7, 7

Suppose that when the following IF statement is executed, $U = 4.$, $V = 8.$, and $W = 6.$:

IF (U*V − 7.*W) 17, 38, 2

Since the expression has the value $4(8) - 7(6) = -10 < 0$, control will be transferred to statement number 17.

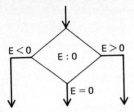

Fig. 4.8   IF Statement Block

Figure 4.8 shows a typical block diagram representation for an IF state-ment. It contains the diamond symbol with one arrow indicating the flow into the statement. Three output paths are shown indicating the flow to be followed depending upon the value of the expression E. The arrow on the right will be followed if the value of E is positive, etc.

We now present several illustrations of the use of the IF (and GO TO) statements in programs; each program will be preceded by a flowchart.

Illustration: Write a complete Fortran program to read a card containing the values of the variables A, B, C, and P, punched 4F8.0. If $b - 5ac$ is negative, compute $q = 7p^3$; if $b - 5ac$ is positive or zero, compute $q = 5p^4$. In either case, print q.

Fig. 4.9

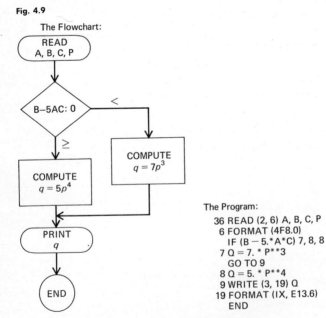

The Flowchart:

The Program:
```
36 READ (2, 6) A, B, C, P
 6 FORMAT (4F8.0)
   IF (B − 5.*A*C) 7, 8, 8
 7 Q = 7. * P**3
   GO TO 9
 8 Q = 5. * P**4
 9 WRITE (3, 19) Q
19 FORMAT (IX, E13.6)
   END
```

A little reflection (and a perusal of the flowchart in Fig. 4.9) reveals the importance of the GO TO 9 statement in this program. Suppose that in fact $b - 5ac$ is negative. Then, following the direction of the IF statement, statement 7 will be executed calculating the value of q as $7p^3$. Now suppose that the statement GO TO 9 were not present. The next statement executed would be, in sequential order, $q = 5p^4$. This value of q would replace the previously calculated value, and this new value would then be printed. This is not what we want. The GO TO 9 statement is placed after statement 7 in order to bypass statement 8 and go directly to the printing of q calculated from $7p^3$. Note that if $b - 5ac$ is positive or zero, statement 8 calculates q as $5p^4$, and the sequential order of statements immediately causes this value to be printed, just as planned.

Illustration: Write a complete Fortran program to read, one at a time, nine cards each containing values of X, Y, Z, and N, punched 3F10.0, I3. After reading the card, if n is greater than 2, compute $s = (x + y + z)/3$, print s, and return to read another card; if n is less than 2, compute $s = \sqrt{x^2 + y^2 + z^2}$, print s, and return to read another card; if n equals 2, do not compute anything, just return to read another card. After the nine cards have been read, print the number 40 and stop.

The use of a counter to keep track of the number of cards read is a feature of this program. We discuss this process. First, the value of the variable I (the counter variable) is set equal to 1. A card is read, and the value of S is computed or not, depending on the value of N. Statement 3 then advances the counter by 1 since $I + 1 = 2$. The next IF statement checks whether I is more than 9. Since it is not, the IF statement transfers control to statement 7. With I now 2, a *second* card is read. The same internal computing process is again followed, and it is repeated over and over again. However, each time through the value of the counter is checked. When I has advanced to 9, the IF statement following statement 3 again transfers control to the READ statement and the *ninth* card is read. When statement 3 is encountered again, I is advanced to 10. Now the IF statement transfers control to statement 8 since 10 is greater than 9. And, after the printing of the number 40, the program ends. This example is important to understand because it demonstrates the programming techniques of initializing, incrementing, and testing the value of a variable. Review Fig. 4.10 for your information.

The following final example again uses these techniques and introduces a new Fortran statement, CONTINUE.

Illustration: Write a complete Fortran program to evaluate the polynomial $x^3 + 3x^2 - 7x + 10$ for values of x from 1 to 5 in increments of 0.05, that is, for $x = 1.00, 1.05, 1.10, 1.15, \ldots, 4.95, 5.00$. Print each x and the corresponding value of the polynomial. A look at Fig. 4.11 will help you.

The Flowchart:

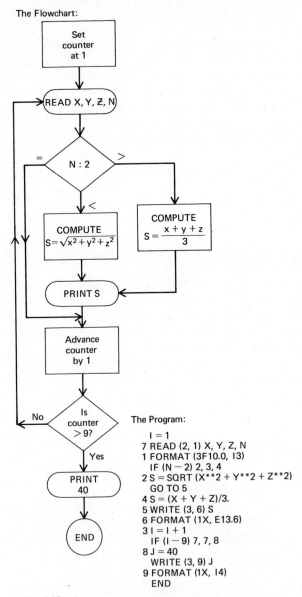

Set counter at 1

READ X, Y, Z, N

N : 2

COMPUTE
$S = \sqrt{x^2 + y^2 + z^2}$

COMPUTE
$S = \dfrac{x + y + z}{3}$

PRINT S

Advance counter by 1

Is counter > 9?

No

Yes

PRINT 40

END

The Program:

```
  I = 1
7 READ (2, 1) X, Y, Z, N
1 FORMAT (3F10.0, I3)
  IF (N − 2) 2, 3, 4
2 S = SQRT (X**2 + Y**2 + Z**2)
  GO TO 5
4 S = (X + Y + Z)/3.
5 WRITE (3, 6) S
6 FORMAT (1X, E13.6)
3 I = I + 1
  IF (I − 9) 7, 7, 8
8 J = 40
  WRITE (3, 9) J
9 FORMAT (1X, I4)
  END
```

Fig. 4.10

The Flowchart:

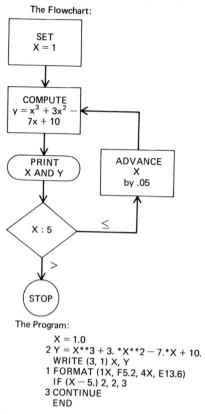

The Program:

```
    X = 1.0
2 Y = X**3 + 3. *X**2 − 7.*X + 10.
    WRITE (3, 1) X, Y
1 FORMAT (1X, F5.2, 4X, E13.6)
    IF (X − 5.) 2, 2, 3
3 CONTINUE
    END
```

Fig. 4.11

CONTINUE is a dummy statement which does not cause anything to be computed. It is used here in order to provide some place to which to transfer control when x is greater than 5. Recall that we have stipulated that control cannot be transferred to the END statement.

EXERCISES

For each of the following exercises construct a flowchart and write the program itself on Fortran coding sheets.

4.15 Write a complete Fortran program to read a card containing the values of the variables B and X (punched 2F10.0), compute $y = bx^2$ and print y; read another card containing another set of values for b and x, compute $y = bx^4$, and print y; read a third card, compute and print $y = bx^6$; continue in the same manner for 12 cards, finally computing and printing $y = bx^{24}$ and printing y.

4.16 Write a complete Fortran program to solve the set of equations

$$\begin{cases} ax + by = c \\ dx + ey = f \end{cases}$$

A deck of 15 cards each containing a, b, c, d, e, and f (punched 6F10.0) is provided. Compute and print x and y for each set of constants. (Refer to Example 4.3.)

4.17 Write a complete Fortran program to read a card containing the values of the variables A, B, C, and X (punched 4F6.0). If $C > A + 3B$, compute and print $y = x^3$; if $C = A + 3B$, compute and print $y = x^4$; if $C < A + 3B$, compute and print $y = x^5$. In any case, return to read a new card with (new) values of the variables A, B, C, and X.

4.18 Write a complete Fortran program to read a card with the numbers, all positive, A, B, C, and D, punched 4F10.0. Determine and print the value of the variable BIG if BIG is the largest of A, B, C, and D.

4.19 Write a complete Fortran program to read a card with the positive numbers A, B, C, and D, punched 4F10.0, read another card with the value of N, punched I2. Let BIG be the largest of A, B, C, and D and SMALL be the smallest of A, B, C, and D. If $N = 0$, determine and print BIG; if $N \neq 0$, determine and print SMALL.

4.20 Write a complete Fortran program to read three cards containing, one per card, the values of XFST, DELX, XLST, each punched F8.3. Compute $z = (4 + x^3)^{1/3} + 5.6x^2$ for values of x from XFST to XLST in increments of DELX. Print each x and the corresponding value of z.

4.21 Write a complete Fortran program to read a card with the values of N, R, and S, punched I4, 2F10.0. If $n = 1$ and $r < s$, compute $z = 4r - 3s$ and print z; if $n = 1$ and $r \geq s$, compute $z = 3s - 4r$ and print z; if $n \neq 1$, print the number 567.

4.22 Write a complete Fortran program to read a card with the value of $k > 0$ and A, C, and X (I4, 3F10.0). If $k = 1$, compute $d =$

$(a - c)^3 + x^2$ and print d; if $k = 2$, compute $d = (a + c^2 + x^3)^2$ and print k and d; if $k > 2$, compute $d = (a + c + x)^{3/2}$ and print k and d. In any case, return to read a new card.

4.23 Write a complete Fortran program to read a card containing values of M and N (punched 2I5) and determine whether M is divisible by N. If M is divisible by N, print the number 5; if it is not, print the number 7. Assume that M is greater than N.

5

Arrays, Matrices, Determinants

5.1 ONE–DIMENSIONAL ARRAYS

A set of singly subscripted variables like $\{x_1, x_2, x_3, x_4, \ldots, x_n\}$ is called a *one-dimensional array*. Each variable uses the common array name x and has one integer subscript. In Fortran these subscripted variables are written X(1), X(2), X(3), X(4), . . . , X(N).

Recall the expression

$$\sum_{i=1}^{n} x_i$$

which means that the variables $x_1, x_2, \ldots, x_n$ are to be added. Let us now consider how we could find such a sum as part of a Fortran program. Suppose that the numerical values of the 80 variables X(1), X(2), . . . , X(80) are known to the computer and we want to calculate the sum of these 80 values. Consider first the following set of instructions:

    I = 1
    SUM = 0.
    SUM = SUM + X(I)

The first two statements set the values of I and of SUM initially. The third statement's righthand side is now evaluated. Thus, to SUM (now 0) is added X(I), that is, X(1) since I is now 1. Then this result is stored at SUM. Now SUM contains the value of X(1). The first element of the array X has thus been stored in SUM. Now suppose we extend the set of statements above to

```
       I = 1
       SUM = 0.
   16  SUM = SUM + X(I)
       I = I + 1
       GO TO 16
```

The first three statements are as before. Then the next two statements say: Add 1 to I (now I = 2) and GO TO 16. Statement 16 says: Add to SUM [which now contains X(1)] the value of X(I) [that is, X(2) since now I = 2]; store the result in SUM. It is clear that now SUM has stored in it the sum of X(1) and X(2). Then the process will be repeated; 1 will be added to I (now I = 3), GO TO 16, etc. A nice loop has been set up. However, this process cannot be permitted to go on indefinitely because X(80) is the last element of the array which we desire to add on. This situation is easily taken care of by telling the computer to stop repeating the process as soon as I becomes 81. We thus extend the previous partial program to the following:

```
       I = 1
       SUM = 0.
   16  SUM = SUM + X(I)
       I = I + 1
       IF (I − 80) 16, 16, 18
   18  WRITE (3, 19) SUM
   19  FORMAT (1X, E13.6)
```

As you can see, when the IF statement tests the value of I and finds it is 81, SUM will contain the sum of the 80 elements of the array X. The partial program then calls for this value to be printed.

We have assumed throughout this last discussion that somehow the 80 values of the elements of the array had been stored and thus were known to the computer. We now present a method to read in and store the elements of a one-dimensional array. First, the computer must know, before it encounters arrays in the body of a program, the name and number of elements of the arrays so that it can reserve in advance the storage positions required. We inform the computer of this by using a DIMENSION statement. The DIMENSION statement identifies the names of the arrays and indicates the maximum number of elements of each array. It must occur in the program before the subscripted variables whose dimensions it specifies appear in the program. If a program contains subscripted variables, the DIMENSION statement containing the required information is usually placed first in the program. For one-dimensional arrays the DIMENSION statement has the form shown in the following example:

DIMENSION X(80), ZUTZ(23), NN(6)

It specifies to the computer during compilation that three subscripted one-dimensional variables are used later in the program, that the array X has at most 80 elements, the array ZUTZ has at most 23 elements, and the array NN has at most six elements. Note that arrays can have real or integer variable names. Clearly, the array NN will have integer values for the elements of the array. Note that the number of elements specified in each case must be an actual constant, it *cannot* be a variable. Any number of arrays (within reason) can be specified in a single DIMENSION statement.

We now present a complete Fortran program to read in the elements of an 80-element array X, sum its elements, and output this sum. We know that a little study on your part of the flowchart and program shown in Fig. 5.1 will enable you to write similar programs.

EXERCISES

5.1 Write a complete Fortran program to read and store the 30 elements of the one-dimensional array X, punched one per card F10.0, 30 cards in all. Compute and print the sum of the squares of the elements

$$\sum_{i=1}^{30} x_i^2$$

5.2 Write a complete Fortran program to read in the 40 elements of the one-dimensional array A, 40 cards, one element per card, punched F6.0. Then read a card with the value of M. If M is less than 1 *or* greater than 40, print the value of M. If $1 \leq M \leq 40$, compute and print the sum of the first M elements of the array.

5.3 Write a complete Fortran program to read in the 50 elements of two one-dimensional arrays X and Y. The elements of the arrays are punched one value of the X array and one value of the Y array per card, 50 cards in all, (x_1,y_1) on the first card, (x_2,y_2) on the second card, etc. Each element is punched F10.0. Compute and print

The Flowchart:

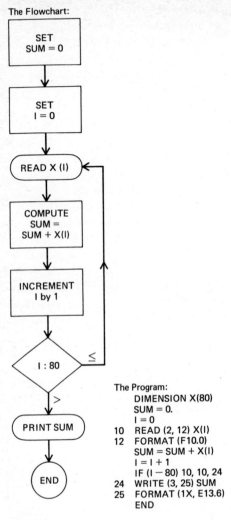

The Program:
```
      DIMENSION X(80)
      SUM = 0.
      I = 0
10    READ (2, 12) X(I)
12    FORMAT (F10.0)
      SUM = SUM + X(I)
      I = I + 1
      IF (I − 80) 10, 10, 24
24    WRITE (3, 25) SUM
25    FORMAT (1X, E13.6)
      END
```

Fig. 5.1

$$c = \sum_{j=1}^{50} (3x_j + 4y_j)^3$$

5.4 Write a complete Fortran program to read 40 cards. The first 10 cards contain the 20 elements of the one-dimensional array A, *two elements per card*, punched 2F10.0. The next 30 cards contain the 30 elements of the one-dimensional array B, punched one element per card, F10.0. Compute

$$X = \sum_{i=1}^{20} a_i + \sum_{i=1}^{30} b_i^2$$

Print x.

5.5 Write a complete Fortran program to read in the 24 elements of the one-dimensional array D, punched one per card F7.0. Compute

$$x = \sum_{i=1}^{24} (d_i^2 + d_i + 5)$$

Print the elements of the array D, one per line, then print the value of x.

5.6 Write a complete Fortran program to read in the 40 elements of the one-dimensional array G, 10 cards each containing four elements, punched 4F10.0. Compute

$$x = \left(\sum_{i=1}^{40} g_i^2 \right)^{1/2}$$

Print x.

5.2 MATRICES

5.2.1 Introduction. Definitions

The set of subscripted variables $\{x_{1,1}, x_{2,1}, x_{1,2}, x_{2,2}, x_{3,2}, \ldots\}$ is called a *two-dimensional array*; each variable uses the common array name x and *two* integer subscripts. In Fortran, these variables are written as X(1,1), X(2,1), X(1,2), X(2,2), X(3,2), ... Variables with two subscripts are usually placed in geometric rectangular arrangements called *matrices*. For example,

$$\begin{bmatrix} x_{1,1} & x_{1,2} & x_{1,3} \\ x_{2,1} & x_{2,2} & x_{2,3} \\ x_{3,1} & x_{3,2} & x_{3,3} \\ x_{4,1} & x_{4,2} & x_{4,3} \end{bmatrix}$$

is a 4×3 matrix; it has four rows and three colunns. Note that the first subscript indicates the number of the row in which the element occurs, the second subscript indicating the number of the column in which the element occurs.

In general, the $m \times n$ matrix $\mathbf{A}$ is the array

$$\begin{bmatrix} a_{1,1} & a_{1,2} & a_{1,3} & \cdots & a_{1,n} \\ a_{2,1} & a_{2,2} & a_{2,3} & \cdots & a_{2,n} \\ \cdot & \cdot & \cdot & \cdot & \cdot \\ a_{m,1} & a_{m,2} & a_{m,3} & \cdots & a_{m,n} \end{bmatrix}$$

A matrix may be $1 \times n$ and thus have 1 row and n columns; it is then called, appropriately enough, a *row matrix*. Similarly, an $n \times 1$ matrix has n rows and 1 column and is called a *column matrix*. A matrix is *square* if $m = n$, that is, if it has the same number of rows as columns. Thus, $\begin{bmatrix} 4 & 2 \\ 5 & -1 \end{bmatrix}$ is a 2×2 (square) matrix. The set of elements of a square matrix which lie along the diagonal extending from the upper left element to the lower right element is called the *principal diagonal* of the matrix. The elements 3, 4, 5 of the matrix

$$\begin{bmatrix} 3 & 2 & -5 \\ 0 & 4 & 7 \\ 2 & -9 & 5 \end{bmatrix}$$

constitute its principal diagonal. A square matrix with all its elements *below* the principal diagonal equal to zero is called *upper triangular*. Thus, the matrix

$$\begin{bmatrix} 3 & 4 & 5 & 6 \\ 0 & 2 & 3 & -7 \\ 0 & 0 & 4 & 12 \\ 0 & 0 & 0 & 5 \end{bmatrix}$$

is upper triangular. Similarly, a matrix with all its elements *above* the principal diagonal equal to zero is called *lower triangular*. A matrix which is both lower and upper triangular is called a *diagonal* matrix. For example,

$$\begin{bmatrix} 3 & 0 & 0 & 0 \\ 0 & 4 & 0 & 0 \\ 0 & 0 & 6 & 0 \\ 0 & 0 & 0 & -7 \end{bmatrix}$$

is diagonal. Finally, a diagonal matrix in which all the elements along the principal diagonal are 1s is called a *unit* matrix. Thus, the 3×3 unit matrix is

$$\begin{bmatrix} 1 & 0 & 0 \\ 0 & 1 & 0 \\ 0 & 0 & 1 \end{bmatrix}$$

A matrix all of whose elements are equal to zero is called a *zero matrix*. $\begin{bmatrix} 0 & 0 & 0 \\ 0 & 0 & 0 \end{bmatrix}$ is the 2×3 zero matrix.

As you have noted above, we often give a matrix a single letter name, usually the common variable name, like **A**. We also write $\mathbf{A} = [a_{i,j}]$ for shorthand. We write equations concerning matrices at two levels. On one level is a *matrix* equation in which each letter represents a matrix and the operation symbols are *matrix operations,* such as $\mathbf{A} + 3\mathbf{B} = \mathbf{C} \times \mathbf{D}$. On the other level we can write a *numerical* equation involving the elements of the matrices, as $c_{i,j} = b_{i,j}{}^2 + 6$, where $c_{i,j}$ and $b_{i,j}$ represent elements of the matrices **C** and **B**, respectively. For example, as we shall soon see in detail, when one is asked to multiply the matrices **A** and **B**, that is, find $\mathbf{A} \times \mathbf{B}$, the actual calculations consist of ordinary arithmetic operations on the elements of the matrices **A** and **B**.

5.2.2 Algebra of Matrices

Two matrices are *equal* if they are identical, that is, if they have the same number of rows and the same number of columns and if each element of one is equal numerically to the corresponding element of the other.

We shall first define the operations of addition and subtraction of two matrices. We write $\mathbf{C} = \mathbf{A} \pm \mathbf{B}$, and the elements of **C** are defined by $c_{i,j} = a_{i,j} \pm b_{i,j}$. In order that two matrices may be added or subtracted, they must have the same number of rows and the same number of columns. The sum matrix **C** then has for its elements simply the sum (or difference) of the corresponding elements of **A** and **B**. Obviously, the order of **C** will be the same as the common order of **A** and **B**. Here are some examples. Suppose

$$\mathbf{A} = \begin{bmatrix} 2 & 3 & 4 \\ 3 & -5 & 6 \end{bmatrix}$$

and

$$\mathbf{B} = \begin{bmatrix} 0 & 4 & 15 \\ 4 & 5 & -3 \end{bmatrix}$$

Then

$$\mathbf{C} = \mathbf{A} + \mathbf{B} = \begin{bmatrix} 2 & 7 & 19 \\ 7 & 0 & 3 \end{bmatrix}$$

and

$$D = A - B = \begin{bmatrix} 2 & -1 & -11 \\ -1 & -10 & 9 \end{bmatrix}$$

Clearly, the operation of matrix addition is commutative, that is, $A + B = B + A$; the order of adding is immaterial. Next, we define the *scalar multiple* of a matrix. If a is a number, then by aX we mean the matrix Y such that $y_{i,j} = ax_{i,j}$. *Each* element of X is simply multiplied by the number a. Thus, if

$$X = \begin{bmatrix} 3 & 4 \\ 2 & 5 \\ 0 & -3 \end{bmatrix}$$

then

$$5X = \begin{bmatrix} 15 & 20 \\ 10 & 25 \\ 0 & -15 \end{bmatrix}$$

We can now form linear combinations of matrices, like $A - 3B + 4C$, presuming, of course, that A, B, and C are all of the same size.

The operation of matrix multiplication is rather peculiar and at first glance seems involved. If C is the product of two matrices A and B, we write $C = A \times B$. We first point out that a certain kind of compatibility must be satisfied by the matrices A and B if they are to be multiplied. The number of *columns* of A must be the same as the number of *rows* of B. Thus, if A is an $i \times j$ matrix, then B must be a $j \times k$ matrix. The product matrix C will then be an $i \times k$ matrix. Thus, if A is 3×5 and B is 5×7, then C will be 3×7. Two matrices which do not satisfy this criterion cannot be multiplied.

The operation of matrix multiplication is not commutative. In general, $A \times B$ is not equal to $B \times A$. One must therefore keep good track of the order of the multiplication. In the product $A \times B$ we often say that A *premultiplies* B, or equivalently that B *postmultiplies* A.

Suppose we have the $m \times p$ matrix A and the $p \times n$ matrix B. The elements of the product matrix $C = A \times B$ are found in the following way: the elements $c_{i,j}$ are computed by multiplying the elements of the ith row of A by the elements of the jth column of B and adding these products. That is,

$$c_{i,j} = \sum_{k=1}^{p} a_{i,k} b_{k,j}$$

For instance, if **A** is 3×4 and **B** is 4×5, then

$$c_{2,3} = \sum_{k=1}^{4} a_{2,k} b_{k,3}$$
$$= a_{2,1}b_{1,3} + a_{2,2}b_{2,3} + a_{2,3}b_{3,3} + a_{2,4}b_{4,3}$$

Note, again, that $c_{2,3}$ is found by summing the products of the elements of the second row of **A** by the third column of **B**. Similarly, $c_{1,5}$ is found by summing the products of the elements of the first row of **A** by the fifth column of **B**. It is clear now why, to multiply the matrices **A** and **B**, the number of columns of **A** must equal the number of rows of **B**. If this were not so, the required products of elements could not be found.

Example. If

$$\mathbf{A} = \begin{bmatrix} 2 & 3 & 4 \\ 3 & 5 & 2 \end{bmatrix} \qquad \mathbf{B} = \begin{bmatrix} 25 & 1 \\ 0 & 2 \\ 3 & 4 \end{bmatrix}$$

then

$$\mathbf{C} = \mathbf{A} \times \mathbf{B}$$
$$= \begin{bmatrix} 2 & 3 & 4 \\ 3 & 5 & 2 \end{bmatrix} \times \begin{bmatrix} 25 & 1 \\ 0 & 2 \\ 3 & 4 \end{bmatrix}$$
$$= \begin{bmatrix} 50 + 0 + 12 & 2 + 6 + 16 \\ 75 + 0 + 6 & 3 + 10 + 8 \end{bmatrix}$$
$$= \begin{bmatrix} 62 & 24 \\ 81 & 21 \end{bmatrix}$$

If the elements are simple numbers like these, one usually performs the products and sums in one's head and simply enters the result in the proper position in the product matrix. A two-finger method can be used. Run your left index finger across the first row of **A** and at the same time run your right index finger down the first column of **B**, multiply corresponding elements, and add as you go. The result will be the element $c_{1,1}$ in the product matrix **C**; place it in the proper place in **C**. Then again run your fingers across the first row of **A** and down the second column of **B**, multiply and add as you go. Place this result in the position $c_{1,2}$ in **C**. Continue in the same way across each row of **A** and down each column of **B**, being sure to place the respective results in the proper place in **C**.

If **A** is a square matrix and **U** is the unit matrix of the same size as **A**, then one can easily verify that $\mathbf{A} \times \mathbf{U} = \mathbf{U} \times \mathbf{A} = \mathbf{A}$. That is, multiplying by

U leaves **A** unchanged, and, in addition, this multiplication is commutative. In the previous example of multiplication of matrices, **A** was 2×3, **B** was 3×2, and the product **C** was, of course, 2×2. Clearly, the product **B** $\times$ **A** is also possible in this case. Thus,

$$\mathbf{B} \times \mathbf{A} = \begin{bmatrix} 25 & 1 \\ 0 & 2 \\ 3 & 4 \end{bmatrix} \times \begin{bmatrix} 2 & 3 & 4 \\ 3 & 5 & 2 \end{bmatrix}$$

$$= \begin{bmatrix} 53 & 80 & 102 \\ 6 & 10 & 4 \\ 18 & 29 & 20 \end{bmatrix}$$

Now **B** $\times$ **A** is 3×3, and clearly **A** $\times$ **B** $\neq$ **B** $\times$ **A**.

Example. Let $\mathbf{A} = \begin{bmatrix} 3 & 5 \\ 2 & -3 \end{bmatrix}$, $\mathbf{B} = \begin{bmatrix} 2 & 7 \\ 1 & 4 \end{bmatrix}$; then $\mathbf{A} \times \mathbf{B} = \begin{bmatrix} 11 & 41 \\ 1 & 2 \end{bmatrix}$ while $\mathbf{B} \times \mathbf{A} = \begin{bmatrix} 20 & -11 \\ 11 & -7 \end{bmatrix}$. Again, **A** $\times$ **B** $\neq$ **B** $\times$ **A**.

EXERCISES

5.7 Given $\mathbf{A} = \begin{bmatrix} 3 & 6 \\ 2 & 5 \end{bmatrix}$, $\mathbf{B} = \begin{bmatrix} -3 & 4 \\ 7 & 2 \end{bmatrix}$,

(a) find $\mathbf{C} = \mathbf{A} \times \mathbf{B}$ and $\mathbf{D} = \mathbf{B} \times \mathbf{A}$;
(b) find $\mathbf{E} = 3\mathbf{A} - 5\mathbf{B} + 4\mathbf{U}$;
(c) if $\mathbf{A}^2 = \mathbf{A} \times \mathbf{A}$, $\mathbf{B}^2 = \mathbf{B} \times \mathbf{B}$, find $\mathbf{A}^2 - \mathbf{B}^2$;
(d) find $(\mathbf{A} + \mathbf{B}) \times (\mathbf{A} - \mathbf{B})$;
(e) find $5\mathbf{A}^2 + 4\mathbf{AB} + \mathbf{B}^2$.

5.8 Given

$$\mathbf{A} = \begin{bmatrix} 3 \\ 1 \\ 2 \end{bmatrix}$$

$$\mathbf{B} = \begin{bmatrix} 1 & 2 & 5 & 4 \end{bmatrix}$$

$$\mathbf{C} = \begin{bmatrix} 1 & -2 & 1 \\ 4 & 1 & 5 \\ 0 & 2 & 6 \end{bmatrix}$$

$$\mathbf{D} = \begin{bmatrix} 1 & 5 & -2 \\ -1 & 3 & 6 \end{bmatrix}$$

$$E = \begin{bmatrix} 1 & 1 \\ 2 & 2 \\ 3 & 4 \\ 2 & 5 \end{bmatrix}$$

find the products $D \times C \times A$, $A \times B$, $B \times E$, $E \times D \times C$.

5.2.3 Solution of Simultaneous Linear Equations

We return once again to the problem of solving sets of simultaneous linear equations and present some ways in which matrices and matrix methods can be used. Suppose we have the set of equations:

$$\begin{cases} 3x_1 + 2x_2 + 3x_3 - x_4 = 7 \\ x_1 \quad\quad + x_3 + 7x_4 = 27 \\ 2x_1 + x_2 - 4x_3 \quad\quad = -3 \\ 4x_1 + x_2 + 8x_3 + x_4 = 38 \end{cases} \quad\quad (5.1)$$

The matrix

$$A = \begin{bmatrix} 3 & 2 & 3 & -1 \\ 1 & 0 & 1 & 7 \\ 2 & 1 & -4 & 0 \\ 4 & 1 & 8 & 1 \end{bmatrix}$$

of the coefficients of the four variables is called the *coefficient matrix*. Now consider this *matrix equation:*

$$\begin{bmatrix} 3 & 2 & 3 & -1 \\ 1 & 0 & 1 & 7 \\ 2 & 1 & -4 & 0 \\ 4 & 1 & 8 & 1 \end{bmatrix} \times \begin{bmatrix} x_1 \\ x_2 \\ x_3 \\ x_4 \end{bmatrix} = \begin{bmatrix} 7 \\ 27 \\ -3 \\ 38 \end{bmatrix} \quad\quad (5.2)$$

The matrix equation (5.2) is equivalent to the original set of equations (5.1). To see this, we multiply the left-side matrices. The product is the *column* matrix

$$\begin{bmatrix} 3x_1 + 2x_2 + 3x_3 - x_4 \\ x_1 \quad\quad + x_3 + 7x_4 \\ 2x_1 + x_2 - 4x_3 \\ 4x_1 + x_2 + 8x_3 + x_4 \end{bmatrix}$$

Since this matrix is equal to the column matrix

$$\begin{bmatrix} 7 \\ 27 \\ -3 \\ 38 \end{bmatrix}$$

their corresponding elements must be equal. If you write all this down, you will clearly have just the original set of linear equations (5.1). Now since the matrix equation (5.2) is equivalent to the set of equations (5.1), we wonder whether we can solve the set of equations written in the matrix form (5.2) using only matrix methods. Let us generalize (5.2) to the following form:

$$AX = B \tag{5.3}$$

where A is the coefficient matrix, X is the column matrix of unknowns whose values we are to find in order to solve the set of equations, and B is the column matrix of the righthand members of the equations.

Suppose we could find a matrix, say C, which would have the remarkable property that $C \times A = U$, where U is a unit matrix. We could then *premultiply* each side of Eq. (5.3) by C and get

$$C(AX) = CB$$

Then $(CA)X = CB$ (since matrix multiplication is associative), $UX = CB$ (since $CA = U$), and $X = CB$ (since $UX = X$, as you recall). Thus, according to the matrix equation $X = CB$, CB is the column matrix whose elements are precisely the same as those in the corresponding positions in X. By equating these corresponding elements, we shall then have solved for $x_1, x_2, \ldots , x_n$.

This matrix method looks very promising. The difficulty involved here, though, is that we must first find this peculiar matrix C. The matrix C (if it exists) is called the *inverse* of A. If is, of course, unique. To indicate its defining property with reference to A, we usually write it as A^{-1} (which we read "A inverse"). Thus, $A^{-1} \times A = U$. It turns out also that $A \times A^{-1} = U$. Unfortunately, it turns out that for even relatively small matrices like 4×4 it is rather difficult to find the inverse. We can, by turning the tables and solving simultaneous equations by the ordinary algebraic method, easily find the inverse of a 2×2 matrix. Suppose that we have the matrix $A = \begin{bmatrix} a & b \\ c & d \end{bmatrix}$, $a, b, c,$ and d known.

Let its inverse A^{-1} be $\begin{bmatrix} e & f \\ g & h \end{bmatrix}$, $e, f, g,$ and h unknown and to be found. Then, since $A \times A^{-1}$ must equal U, we have

$$\begin{bmatrix} a & b \\ c & d \end{bmatrix} \times \begin{bmatrix} e & f \\ g & h \end{bmatrix} = \begin{bmatrix} 1 & 0 \\ 0 & 1 \end{bmatrix}$$

Multiplying, we have

$$\begin{bmatrix} ae + bg & af + bh \\ ce + dg & cf + dh \end{bmatrix} = \begin{bmatrix} 1 & 0 \\ 0 & 1 \end{bmatrix}$$

This leads to the four simultaneous equations

$$\begin{cases} ae + bg = 1 \\ af + bh = 0 \\ ce + dg = 0 \\ cf + dh = 1 \end{cases} \tag{5.4}$$

Solving the second equation for f, we have $f = -bh/a$. Substituting this for f in the fourth equation, we have $(-bc/a)h + dh = 1$, or $h(d - bc/a) = 1$ and

$$h = \frac{1}{d - bc/a}$$
$$= \frac{a}{ad - bc}$$

We also have $f = -bh/a = -b/(ad - bc)$. Again, solving the third equation for e, we have $e = -dg/c$. Substituting into the first equation, we have

$$-\frac{ad}{c}g + bg = 1 \qquad g\left(b - \frac{ad}{c}\right) = 1 \qquad g = \frac{1}{b - \dfrac{ad}{c}} \quad = \quad \frac{c}{bc - ad} \quad = \quad \frac{-c}{ad - bc}$$

And, since $e = -dg/c$, $e = d/(ad - bc)$. Thus,

$$\mathbf{A}^{-1} = \begin{bmatrix} \dfrac{d}{ad - bc} & \dfrac{-b}{ad - bc} \\ \dfrac{-c}{ad - bc} & \dfrac{a}{ad - bc} \end{bmatrix} = \frac{1}{ad - bc}\begin{bmatrix} d & -b \\ -c & a \end{bmatrix}$$

A comparison of $\mathbf{A}^{-1}$ and $\mathbf{A}$ reveals the following little rule: $\mathbf{A}^{-1}$ is found by interchanging a and d, changing the signs of b and c, and dividing by the quantity $ad - bc$.

Clearly, $ad - bc$ cannot equal zero. $\mathbf{A}$ will not have an inverse (and will be called *singular*) if $ad - bc = 0$. The matrix $\begin{bmatrix} 4 & 8 \\ 1 & 2 \end{bmatrix}$ does not have an inverse.

Example. If

$$\mathbf{A} = \begin{bmatrix} 3 & 4 \\ -2 & 7 \end{bmatrix}$$

then

$$\mathbf{A}^{-1} = \frac{1}{29} \begin{bmatrix} 7 & -4 \\ 2 & 3 \end{bmatrix}$$

$$= \begin{bmatrix} 7/29 & -4/29 \\ 2/29 & 3/29 \end{bmatrix}$$

You can easily verify that

$$\mathbf{A} \times \mathbf{A}^{-1} = \begin{bmatrix} 1 & 0 \\ 0 & 1 \end{bmatrix}$$

Example. Solve

$$\begin{cases} 3x + 4y = 7 \\ -2x + 7y = 12 \end{cases}$$

using matrix methods. We have

$$\begin{bmatrix} 3 & 4 \\ -2 & 7 \end{bmatrix} \begin{bmatrix} x \\ y \end{bmatrix} = \begin{bmatrix} 7 \\ 12 \end{bmatrix}$$

Thus,

$$\begin{bmatrix} x \\ y \end{bmatrix} = \begin{bmatrix} 3 & 4 \\ -2 & 7 \end{bmatrix}^{-1} \begin{bmatrix} 7 \\ 12 \end{bmatrix}$$

$$= \frac{1}{29} \begin{bmatrix} 7 & -4 \\ 2 & 3 \end{bmatrix} \begin{bmatrix} 7 \\ 12 \end{bmatrix}$$

$$= \frac{1}{29} \begin{bmatrix} 1 \\ 50 \end{bmatrix}$$

$$= \begin{bmatrix} 1/29 \\ 50/29 \end{bmatrix}$$

Therefore, $x = 1/29$ and $y = 50/29$.

5.2.4 Elementary Row Operations. Inverses

Unfortunately, the little rule above for inverting 2×2 matrices cannot be easily extended to cover the cases of matrices of larger size, and so we now present an entirely different matrix technique which experts in numerical analysis use. This new method can solve simultaneous linear equations and produce the inverse of the coefficient matrix at the same time. We must first introduce new notions of elementary row operations on matrices and of equivalent matrices.

The following operations on the *rows* of any matrix **A** result in a new matrix which we say is *equivalent* to the matrix **A**: (1) Multiply each of the elements of any row by a fixed nonzero constant; (2) interchange any two rows; (3) fix any row, then add a fixed multiple of each of its elements to the corresponding elements of any other row.

Clearly, any of these so-called *elementary row operations* on a matrix **A** can result in a matrix **B** which is not *equal* to **A**. But the new matrix **B** is *equivalent* to the original matrix **A**, in a sense which the following applications will make clear. We can apply any of these operations successively on a matrix and thus get a chain of equivalent matrices.

Example. $\begin{bmatrix} 3 & 7 & 1 & 2 \\ 4 & 3 & 1 & 5 \end{bmatrix} \sim \begin{bmatrix} 9 & 21 & 3 & 6 \\ 4 & 3 & 1 & 5 \end{bmatrix} \sim \begin{bmatrix} 4 & 3 & 1 & 5 \\ 9 & 21 & 3 & 6 \end{bmatrix}$

In the first step we have multiplied the first row by 3, and in the second we have interchanged the two rows.

Example. Suppose we have the matrix

$$\begin{bmatrix} 3 & 4 & 1 \\ 5 & 2 & 3 \\ -1 & 2 & 7 \end{bmatrix}$$

and we *fix* the second row, then add two times this row to row 1. We now have

$$\begin{bmatrix} 13 & 8 & 7 \\ 5 & 2 & 3 \\ -1 & 2 & 7 \end{bmatrix}$$

which is equivalent to

$$\begin{bmatrix} 3 & 4 & 1 \\ 5 & 2 & 3 \\ -1 & 2 & 7 \end{bmatrix}$$

Now we present several interesting and useful applications of equivalent matrices. First, suppose a square matrix **A** has an inverse. By an adroit selection of a sequence of elementary row operations, it is possible to transform **A** into a unit matrix which is equivalent to **A**. The fascinating fact we shall make use of is that the *very same sequence* of elementary row operations that transformed **A** into **U** will transform **U** into **A**⁻¹. Here is how this is done in actual practice. Suppose we have the 3 × 3 matrix

$$\mathbf{A} = \begin{bmatrix} 4 & 1 & 3 \\ 4 & 7 & 2 \\ 3 & 1 & -5 \end{bmatrix}$$

We *augment* **A** into the 3 × 6 matrix

$$\begin{bmatrix} 4 & 1 & 3 & 1 & 0 & 0 \\ 4 & 7 & 2 & 0 & 1 & 0 \\ 3 & 1 & -5 & 0 & 0 & 1 \end{bmatrix}$$

by appending on the right of **A** the 3 × 3 unit matrix. Then as you pick out and apply the sequence of elementary row operations that will transform **A** into **U**, instead of applying them only to **A**, you apply them to the entire row of the augmented matrix. So, in effect, you are automatically applying exactly the same sequence of elementary row operations to **U**. Thus, when **A** → **U**, **U** → **A**⁻¹. That is, the 3 × 3 matrix on the right will be exactly the inverse of **A**. We shall have an example of this in a moment.

Now suppose you have a set of linear equations like

$$\begin{cases} 5x + 7y - z = 8 \\ x + 6y + z = 7 \\ 4x - y - 2z = -10 \end{cases}$$

to solve. If you were to go through the standard elimination process for solving them, you would apply such steps as multiplying an equation through by a fixed number, adding one equation to another, and possibly interchanging any two equations. You realize that these steps are just like elementary row operations applied to the entire equation in each case. One could reduce the set of equations to the following set:

$$x = \underline{\qquad}$$
$$y = \underline{\qquad}$$
$$z = \underline{\qquad}$$

from which the solution is obvious. All this leads one to conclude that exactly the same effect can be achieved using only the matrix of the coefficients augmented by the righthand column of numbers and applying elementary row operations until this augmented matrix has the form

$$\begin{bmatrix} 1 & 0 & 0 & a \\ 0 & 1 & 0 & b \\ 0 & 0 & 1 & c \end{bmatrix}$$

The solution of the set of simultaneous linear equations would then be $x = a$, $y = b$, $z = c$.

We are now going to present an extensive example which incorporates both of these techniques. We shall solve a set of linear equations and *at the same time* find the inverse of its coefficient matrix.

Example. Given the set

$$\begin{cases} x + y + z = 2 \\ 2x - y + z = -1 \\ x + 4y - 2z = 11 \end{cases}$$

use the matrix method just described to solve the equations and at the same time find the inverse of the coefficient matrix.

Solution: First form the (very) augmented matrix

$$\begin{bmatrix} 1 & 1 & 1 & | & 2 & | & 1 & 0 & 0 \\ 2 & -1 & 1 & | & -1 & | & 0 & 1 & 0 \\ 1 & 4 & -2 & | & 11 & | & 0 & 0 & 1 \end{bmatrix}$$

by appending to the coefficient matrix both the column matrix of the righthand sides of the equations *and* the 3 × 3 unit matrix. We now proceed to apply a sequence of elementary row operations which will transform **A** into the unit matrix **U**, but we apply each operation to the *entire* row at each step.

Step 1 Fix row 1. Multiply it through by −2 and add to row 2:

$$\begin{bmatrix} 1 & 1 & 1 & 2 & 1 & 0 & 0 \\ 0 & -3 & -1 & -5 & -2 & 1 & 0 \\ 1 & 4 & -2 & 11 & 0 & 0 & 1 \end{bmatrix}$$

Step 2 Fix row 1.   Multiply through by -1 and add to row 3:

$$\begin{bmatrix} 1 & 1 & 1 & 2 & 1 & 0 & 0 \\ 0 & -3 & -1 & -5 & -2 & 1 & 0 \\ 0 & 3 & -3 & 9 & -1 & 0 & 1 \end{bmatrix}$$

Step 3 Multiply row 2 through by $-\frac{1}{3}$:

$$\begin{bmatrix} 1 & 1 & 1 & 2 & 1 & 0 & 0 \\ 0 & 1 & \frac{1}{3} & \frac{5}{3} & \frac{2}{3} & -\frac{1}{3} & 0 \\ 0 & 3 & -3 & 9 & -1 & 0 & 1 \end{bmatrix}$$

Step 4 Fix row 2.   Multiply through by -3 and add to row 3:

$$\begin{bmatrix} 1 & 1 & 1 & 2 & 1 & 0 & 0 \\ 0 & 1 & \frac{1}{3} & \frac{5}{3} & \frac{2}{3} & -\frac{1}{3} & 0 \\ 0 & 0 & -4 & 4 & -3 & 1 & 1 \end{bmatrix}$$

Step 5 Multiply row 3 through by $-\frac{1}{4}$:

$$\begin{bmatrix} 1 & 1 & 1 & 2 & 1 & 0 & 0 \\ 0 & 1 & \frac{1}{3} & \frac{5}{3} & \frac{2}{3} & -\frac{1}{3} & 0 \\ 0 & 0 & 1 & -1 & \frac{3}{4} & -\frac{1}{4} & -\frac{1}{4} \end{bmatrix}$$

Step 6 Fix row 2.   Multiply through by -1 and add to row 1:

$$\begin{bmatrix} 1 & 0 & \frac{2}{3} & \frac{1}{3} & \frac{1}{3} & \frac{1}{3} & 0 \\ 0 & 1 & \frac{1}{3} & \frac{5}{3} & \frac{2}{3} & -\frac{1}{3} & 0 \\ 0 & 0 & 1 & -1 & \frac{3}{4} & -\frac{1}{4} & -\frac{1}{4} \end{bmatrix}$$

Step 7 Fix row 3.   Multiply through by $-\frac{1}{3}$ and add to row 2:

$$\begin{bmatrix} 1 & 0 & \frac{2}{3} & \frac{1}{3} & \frac{1}{3} & \frac{1}{3} & 0 \\ 0 & 1 & 0 & 2 & \frac{5}{12} & -\frac{1}{4} & \frac{1}{12} \\ 0 & 0 & 1 & -1 & \frac{3}{4} & -\frac{1}{4} & -\frac{1}{4} \end{bmatrix}$$

Step 8 Fix row 3.   Multiply through by $-\frac{2}{3}$ and add to row 1:

$$\begin{bmatrix} 1 & 0 & 0 & 1 & -\frac{1}{6} & \frac{1}{2} & \frac{1}{6} \\ 0 & 1 & 0 & 2 & \frac{5}{12} & -\frac{1}{4} & \frac{1}{12} \\ 0 & 0 & 1 & -1 & \frac{3}{4} & -\frac{1}{4} & -\frac{1}{4} \end{bmatrix}$$

A has now been transformed into **U**. The fourth column is thus the column of unknowns. So, $x = 1$, $y = 2$, $z = -1$. The last three columns are the inverse of **A**. Thus,

$$\mathbf{A}^{-1} = \begin{bmatrix} -\frac{1}{6} & \frac{1}{2} & \frac{1}{6} \\ \frac{5}{12} & -\frac{1}{4} & \frac{1}{12} \\ \frac{3}{4} & -\frac{1}{4} & -\frac{1}{4} \end{bmatrix}$$

$$= \frac{1}{12} \begin{bmatrix} -2 & 6 & 2 \\ 5 & -3 & 1 \\ 9 & -3 & -3 \end{bmatrix}$$

A quick check on your part will reveal that $\mathbf{A} \times \mathbf{A}^{-1}$ is in fact **U**, and so what we are calling $\mathbf{A}^{-1}$ here is truly the inverse of **A**.

You should study this example and take note of the systematic way the matrix **A** is transformed into **U**. This method can easily be applied to any number of equations, but as the number gets larger, the manual work, though arithmetically simple, becomes almost tedious even when the numbers in the original matrix are simple integers. If the coefficients are rational numbers in decimal form with a large number of significant figures, then one could not hope to do the operations manually. Obviously, one would seek the help of the computer. We hope that in your later work in computer programming you will be able to write a Fortran program with ease which will solve any reasonable number of simultaneous linear equations using the method of elementary row operations and which will also invert a matrix of not too unreasonable size.

EXERCISES

5.9 Write the following set of simultaneous linear equations in the matrix form $\mathbf{AX} = \mathbf{B}$. Use the method of matrix inverse to solve this matrix equation for **X** and thus solve the set of equations:

$$\begin{cases} 5x + 4y = 7 \\ 2x + 7y = 15 \end{cases}$$

5.10 Use the matrix method of elementary row operations to transform the matrix

$$\begin{bmatrix} 1 & 6 & 1 \\ 2 & 2 & 2 \\ -1 & 1 & 3 \end{bmatrix}$$

into

$$\begin{bmatrix} 2 & 0 & 0 \\ 0 & 3 & 0 \\ 0 & 0 & 4 \end{bmatrix}$$

5.11 Augment the matrix

$$\mathbf{A} = \begin{bmatrix} -1 & 4 & 2 \\ 1 & 1 & 5 \\ 4 & 2 & 2 \end{bmatrix}$$

to

$$\begin{bmatrix} -1 & 4 & 2 & 1 & 0 & 0 \\ 1 & 1 & 5 & 0 & 1 & 0 \\ 4 & 2 & 2 & 0 & 0 & 1 \end{bmatrix}$$

Then use the method of elementary row operations described in the text to find $\mathbf{A}^{-1}$.

5.12 Solve the following set of simultaneous linear equations using the method of matrix inverse:

$$\begin{cases} -x + 4y + 2z = 7 \\ x + y + 5z = 16 \\ 4x + 2y + 2z = 11 \end{cases}$$

Hint: Make use of the matrix $\mathbf{A}^{-1}$ of Exercise 5.11.

5.13 Use the matrix method of elementary row operations described in the text to solve the following sets of simultaneous linear equations:

(a) $\begin{cases} x + 2y + z = 11 \\ 2x - y - z = -5 \\ -2x + 2y + 5z = 7 \end{cases}$

(b) $\begin{cases} 2y + 7z = 12 \\ x + 3z = 5 \\ x + 3y = 8 \end{cases}$

(c) $\begin{cases} x + 2y + 3z + 4w = 30 \\ x - y - z + w = 0 \\ y + z - w = 1 \\ 2x + 2z - 3w = -5 \end{cases}$

Hint: In (a) form the augmented matrix

$$\begin{bmatrix} 1 & 2 & 1 & 11 \\ 2 & -1 & -1 & -5 \\ -2 & 2 & 5 & 7 \end{bmatrix}$$

and transform it by means of elementary row operations into the equivalent matrix

$$\begin{bmatrix} 1 & 0 & 0 & a \\ 0 & 1 & 0 & b \\ 0 & 0 & 1 & c \end{bmatrix}$$

5.3 DETERMINANTS

Associated with any square matrix $\mathbf{A}$ is a number called its *determinant.* (Non-square matrices do not have determinants.) We abbreviate this number det $\mathbf{A}$. We shall soon give some rules for finding this number and then by means of some applications try to present some notions as to the reasons for these rules. The determinant associated with a matrix $\mathbf{A}$ is the square array of the very same elements of the matrix except that the square brackets holding the matrix elements together are changed to straight bars on each side. For example, the 2×2 matrix $\mathbf{A} = \begin{bmatrix} 4 & 7 \\ 3 & 5 \end{bmatrix}$ has determinant det $\mathbf{A} = \begin{vmatrix} 4 & 7 \\ 3 & 5 \end{vmatrix}$.

Note: Determinants can also be thought of as existing completely independently of matrices. If you have encountered determinants before, this is probably the context in which you studied their properties. In any case, associated with each determinant is a number called its *value.* Matrices, of course, do not have value. They are strictly structural entities.

It is very easy to evaluate 2×2 determinants. We follow this rule: Find the product of the elements along the principal diagonal and subtract the product of the elements along the other diagonal. For example,

det $\mathbf{A} = \begin{vmatrix} 4 & 7 \\ 3 & 5 \end{vmatrix} = 4(5) - 7(3) = 20 - 21 = -1$. Here are some more examples:

$\begin{vmatrix} 3 & 4 \\ -7 & -5 \end{vmatrix} = -15 + 28 = 13$ and $\begin{vmatrix} 0 & 4 \\ 2 & 3 \end{vmatrix} = 0 - 8 = -8$.

Evaluation of 3×3 determinants can be done by a slightly fancier rule. Suppose the determinant is

$$\begin{vmatrix} a & b & c \\ d & e & f \\ g & h & i \end{vmatrix}$$

Its value is given by $aei + bfg + cdh - (ceg + afh + bdi)$. Since this form is not easy to remember, the following mechanical procedure achieves the same result. Form an "augmented" array by copying the first two columns and placing them on the right of the original array. Find the product of the elements on all the three-element diagonals from upper left to lower right. Sum these products. Now form the products of the elements of the diagonals from upper right to lower left. Sum these products. Now subtract the second sum from the first sum. This process is easier than it sounds. Suppose we have the determinant

$$\begin{vmatrix} 3 & 1 & 2 \\ 4 & 5 & 6 \\ 1 & 2 & -3 \end{vmatrix}$$

The "augmented" array is

$$\begin{vmatrix} 3 & 1 & 2 & 3 & 1 \\ 4 & 5 & 6 & 4 & 5 \\ 1 & 2 & -3 & 1 & 2 \end{vmatrix}$$

The first sum is clearly $3(5)(-3) + (1)(6)(1) + (2)(4)(2) = -45 + 6 + 16 = -23$. The second sum is $(2)(5)(1) + (3)(6)(2) + (1)(4)(-3) = 10 + 36 - 12 = 34$. The value of the determinant is thus $-23 - 34 = -57$.

Before we turn to higher-order determinants, let us present an introduction into the reason for these peculiar methods for evaluating 2×2 and 3×3 determinants. Consider the set of simultaneous linear equations:

$$\begin{cases} ax + by = c \\ dx + ey = f \end{cases}$$

Let us solve by the standard method of elimination. In order to eliminate y, we multiply the first equation through by e, the second equation through by b:

$$\begin{cases} ae\ x + be\ y = ce \\ bd\ x + be\ y = bf \end{cases}$$

Subtracting, we have $(ae - bd)x = ce - bf$. Thus

$$x = \frac{ce - bf}{ae - bd} = \frac{\begin{vmatrix} c & b \\ f & e \end{vmatrix}}{\begin{vmatrix} a & b \\ d & e \end{vmatrix}}$$

In the last step we have made use of our previous definition of the value of a 2×2 determinant. We note that x is the quotient of two determinants, both of which are easy to write (and to evaluate). The denominator is simply the determinant of the coefficients of x and y in the two equations in exactly the same position as they occur in the equations. The numerator determinant is found by replacing the column of x coefficients by the column of the righthand numbers of the equations. By solving in a similar way, one can easily show that

$$y = \frac{\begin{vmatrix} a & c \\ d & f \end{vmatrix}}{\begin{vmatrix} a & b \\ d & e \end{vmatrix}}$$

y has the same denominator determinant as for x. The numerator is found by replacing the column of y coefficients in this determinant by the column of righthand numbers of the equations. This method of solving simultaneous linear equations by expressing each unknown as the ratio of two determinants is called *Cramer's rule*.

Example. Solve

$$\begin{cases} 3s - 7t = 12 \\ s + 5t = 17 \end{cases}$$

using Cramer's rule.

$$s = \frac{\begin{vmatrix} 12 & -7 \\ 17 & 5 \end{vmatrix}}{\begin{vmatrix} 3 & -7 \\ 1 & 5 \end{vmatrix}}$$

$$= \frac{60 - (-119)}{15 - (-7)}$$

$$= \frac{179}{22}$$

$$t = \frac{\begin{vmatrix} 3 & 12 \\ 1 & 17 \end{vmatrix}}{22}$$

$$= \frac{51 - 12}{22}$$

$$= \frac{39}{22}$$

Cramer's rule can be extended to any number of simultaneous equations, provided, of course, that the method of evaluating the corresponding determinants has been carefully selected so as to produce the correct results!

Example. Solve

$$\begin{cases} x + 3y + 4z = 17 \\ x - 2y + z = 3 \\ y - z = -2 \end{cases}$$

$$x = \frac{\begin{vmatrix} 17 & 3 & 4 \\ 3 & -2 & 1 \\ -2 & 1 & -1 \end{vmatrix}}{\begin{vmatrix} 1 & 3 & 4 \\ 1 & -2 & 1 \\ 0 & 1 & -1 \end{vmatrix}}$$

$$= \frac{34 + 12 - 6 - (16 - 9 + 17)}{2 + 4 + 0 - (-3 + 0 + 1)}$$

$$= \frac{40 - 24}{6 + 2}$$

$$= \frac{16}{8}$$

$$= 2$$

$$y = \frac{\begin{vmatrix} 1 & 17 & 4 \\ 1 & 3 & 1 \\ 0 & -2 & -1 \end{vmatrix}}{8}$$

$$= \frac{-3 - 8 - (-17 - 2)}{8}$$

$$= \frac{-11 + 19}{8}$$

$$= 1$$

$$z = \frac{\begin{vmatrix} 1 & 3 & 17 \\ 1 & -2 & 3 \\ 0 & 1 & -2 \end{vmatrix}}{8}$$

$$= \frac{4 + 17 - (-6 + 3)}{8}$$

$$= \frac{21 + 3}{8}$$

$$= 3$$

A derivation like the one we went through for the general case of two simultaneous linear equations would reveal that our rule for evaluating third-order
determinants was, to say the least, fortunate. The results of the last example
can be checked by substituting into the original equations.

We note in passing that Cramer's rule will be in some trouble if the denominator determinant has value zero. In fact, if this determinant is zero and
any of the numerator determinants is not zero, the set of equations has no
solution. On the other hand, if the denominator determinant is zero and *all*
the numerator determinants are also zero, the set of equations has infinitely
many solutions.

Unfortunately, these simply mechanical rules for evaluating second-
and third-order determinants do not extend to higher orders. In numerical
analysis determinants of higher order are not evaluated by any extension of
mechanical rules, but interestingly enough their values can be found as by-
products of the elementary row operation techniques we presented in the last
section. Cramer's rule is very seldom used in numerical analysis. Sets of
many simultaneous linear equations are solved by the process of elimination,
iteration, or by the matrix methods using elementary row (or column) operations! Cramer's rule is often used by people who have only two or three
simultaneous equations to solve.

EXERCISES

5.14 Evaluate the following determinants:

(a) $\begin{vmatrix} 3 & 6 \\ -2 & 7 \end{vmatrix}$

(b) $\begin{vmatrix} x & 1 \\ 1 & x \end{vmatrix}$

(c) $\begin{vmatrix} 8 & -4 \\ 16 & -8 \end{vmatrix}$

(d) $\begin{vmatrix} b & 4a \\ c & b \end{vmatrix}$

5.15 Evaluate the following determinants:

(a) $\begin{vmatrix} 1 & 2 & 3 \\ 4 & 5 & 6 \\ 7 & 8 & 9 \end{vmatrix}$

(b) $\begin{bmatrix} 1 & x & x^2 \\ 2 & 3 & 4 \\ 2 & x & x^2 \end{bmatrix}$

(c) $\begin{vmatrix} 0.1 & 0.1 & 0.3 \\ 0.4 & 0.5 & -0.4 \\ 0.6 & 0 & -0.2 \end{vmatrix}$

(d) $\begin{vmatrix} 1 & 2 & 3 \\ 2 & 4 & 6 \\ 5 & 6 & 2 \end{vmatrix}$

(e) $\begin{vmatrix} 1 & 2 & 4 \\ 3 & 4 & -5 \\ 4 & 6 & -1 \end{vmatrix}$

5.16 For the following matrices, find det $\mathbf{A}$:

(a) $\mathbf{A} = \begin{bmatrix} 3 & -2 & -6 \\ 0 & 3 & 5 \\ -2 & 7 & 2 \end{bmatrix}$

(b) $\mathbf{A} = \begin{bmatrix} 4 & 5 & 6 \\ 1 & 2 & 3 \\ 6 & 9 & 12 \end{bmatrix}$

5.17 For the following matrices $\mathbf{A}$ and $\mathbf{B}$, find det $\mathbf{A}$, det $\mathbf{B}$, det $(\mathbf{A} + \mathbf{B})$, det $(\mathbf{A} \times \mathbf{B})$, det $\mathbf{A}$ + det $\mathbf{B}$, det $\mathbf{A} \times$ det $\mathbf{B}$:

(a) $\mathbf{A} = \begin{bmatrix} 4 & 5 \\ 5 & -3 \end{bmatrix}$ $\mathbf{B} = \begin{bmatrix} 2 & 6 \\ -3 & 7 \end{bmatrix}$

(b) $\mathbf{A} = \begin{bmatrix} 4 & -1 & 2 \\ 3 & 6 & 0 \\ 2 & -1 & 3 \end{bmatrix}$ $\mathbf{B} = \begin{bmatrix} 5 & 0 & 2 \\ 3 & 2 & 4 \\ -1 & -3 & -1 \end{bmatrix}$

5.18 Use Cramer's rule to solve the following:

(a) $\begin{cases} 3x + 5y = 56 \\ 2x - 7y = -3 \end{cases}$

(b) $\begin{cases} 6s + 5t = 23 \\ 7s + 3t = 26 \end{cases}$

(c) $\begin{cases} x + 2y + 3z = -5 \\ 2x - y - z = 6 \\ 5x + 7y + 4z = 23 \end{cases}$

(d) $\begin{cases} r + 4s - 5t = 0 \\ 3r - 3s + t = -6 \\ 4r + 3s + 5t = 14 \end{cases}$

5.19 Use Cramer's rule to show that

(a) $\begin{cases} 3x + 5y = 7 \\ 6x + 10y = 11 \end{cases}$

has no solutions;

(b) $\begin{cases} 2x - y = 7 \\ 4x - 2y = 14 \end{cases}$

has infinitely many solutions.   Give at least four explicit different solutions.

5.20 If **U** is 2×2 or 3×3, show that det **U** $= 1$.   Show that det **A** is the product of the diagonal elements if **A** is 2×2 or 3×3 and **A** is diagonal.

5.21 Let **A** be any 2×2 matrix which has an inverse.   Prove that

$$\det \mathbf{A}^{-1} = \frac{1}{\det \mathbf{A}}$$

5.22 Let

$$\mathbf{A} = \begin{bmatrix} 3 & 2 & 1 \\ 4 & 1 & 3 \\ -5 & 2 & 6 \end{bmatrix}$$

Find det **A**, det $\mathbf{A}^2$, and det $\mathbf{A}^{-1}$.

6

Polynomial Functions

6.1 INTRODUCTORY REMARKS

We shall call $f(x)$ a *polynomial function* of *degree n* if it has the form $a_n x^n + a_{n-1} x^{n-1} + \cdots + a_2 x^2 + a_1 x + a_0$, where n is a positive integer. In *mathematical* analysis, the coefficients a_i are *real* numbers. But, as you recall, in *numerical* analysis all real numbers must be approximated by a rational number in decimal form found by truncating the decimal form of the real number after a suitable number of significant figures. Thus, in actual practice the coefficients in the polynomial function are decimal rational numbers approximating, if necessary, the real coefficients. For example, the function $\sqrt{2} x^3 + \frac{1}{3} x^2 - 7.209x + 12.0$ would have to be written as $1.414x^3 + 0.3333x^2 - 7.209x + 12.00$ if all the coefficients are to be expressed to four significant figures. Oftentimes, of course, no approximations at all are required, as in $x^4 + 3x^2 - 7$. In any case, however, $a_n \neq 0$.

In this chapter we shall be concerned with the analytic geometry of these functions, that is, the graphing of such equations as $y = x^2 + 7x - 3$, the solution of polynomial equations, like finding the roots of the equation $x^3 + 4x^2 - 7x - 3 = 0$, the fitting of empirical data to polynomial functions, and an introduction to the theory of equations. We shall also present some examples of Fortran programs which are appropriate to solve problems that are not amenable to simple hand calculations. In order to achieve some of these objectives, it is necessary that we first present the notion of complex numbers.

6.2 COMPLEX NUMBERS

The system of complex numbers is the result of a perfectly natural extension of the system of real numbers. Complex numbers were "invented" so that the roots of *any* polynomial equation could be found and appropriately displayed. A simple case is the equation $x^2 + 4 = 0$. Clearly, there is no *real* number whose square is -4. A new quantity i, called the *imaginary unit*, is defined by the property $i^2 = -1$. This unit i is obviously not a real number. Then the roots of $x^2 + 4 = 0$ are $x = \pm 2i$, since $(2i)^2 = 4i^2 = -4$ and also $(-2i)^2 = 4i^2 = -4$.

The unit i is then combined with real numbers a and b to form the general *complex number* $a + bi$. For example, $-3 + 4i$ and $2 + \sqrt{7}i$ are complex numbers. The a in $a + bi$ is called the *real part* of $a + bi$, and we sometimes write $a = \mathscr{R}e(a + bi)$. The b in $a + bi$ is called the *imaginary part* of $a + bi$, and we often write $b = \mathscr{I}m\ (a + bi)$. For example, $\mathscr{R}e(-3 + 7i) = -3$, and $\mathscr{I}m\ (5 - 6i) = -6$. Note that both the real and imaginary parts of a complex number are *real* numbers. If $b = 0$, the complex number $a + bi$ is simply the real number a. Thus, all real numbers are complex numbers; that is, they can all be written in the complex number form $a + 0i$. If $a = 0$ and $b \neq 0$, then the complex number reduces to bi, which is called a *pure imaginary number*.

As you undoubtedly know, the complex numbers can be combined using the operations of addition, subtraction, multiplication, and division. Under each of these operations the system of complex numbers is *closed*; that is, the result of the operation is again a complex number. In fact, not even the more complicated operation of taking roots of complex numbers will lead out of the set of complex numbers. The system of complex numbers together with these operations which we define in the next paragraph is thus the most complete system of numbers we shall ever need in applied mathematics.

The *addition* of complex numbers is defined by

$$(a + bi) + (c + di) = (a + c) + (b + d)i$$

Thus, the sum is a complex number whose real part is the (real) sum of the real parts of the two numbers and whose imaginary part is the (real) sum of the imaginary parts of the two numbers. For example,

$$(3 + 4i) + (6 - 5i) = (3 + 6) + (4 - 5)i$$
$$= 9 - i$$

The definition of subtraction is simply

$$(a + bi) - (c + di) = (a - c) + (b - d)i$$

Multiplication of complex numbers is done following the ordinary rules of algebra with the proviso that whenever i^2 occurs, it is replaced by -1. Thus, in general,

$$(a + bi)(c + di) = ac + bci + adi + bdi^2$$
$$= ac - bd + (bc + ad)i$$

Example

$$(3 - 4i)(-1 + 7i) = -3 + 4i + 21i - 28i^2$$
$$= -3 + 25i + 28$$
$$= 25 + 25i$$

Note that all positive integral powers of i must be one of the numbers $1, -1, i$, or $-i$. In particular, $i^1 = i$, $i^2 = -1$, $i^3 = -i$, $i^4 = 1$. High powers of i can be easily calculated by taking note of the fact that $i^4 = 1$. Thus, $i^{27} = i^{24}i^3 = (i^4)^6 i^3 = i^3 = -i$ and $i^{482} = i^{480}i^2 = -1$.

Associated with each complex number $a + bi$ is its *conjugate* $a - bi$, formed from $a + bi$ by changing the sign of its imaginary part. We write $\overline{a + bi} = a - bi$.

Division is performed by following this rule: Multiply the numerator and the denominator by the conjugate of the denominator. Note that $(c + di)(c - di) = c^2 + cdi - cdi + d^2 = c^2 + d^2$, a real number.

Example

$$\frac{3 + i}{2 - 3i} = \frac{3 + i}{2 - 3i} \frac{2 + 3i}{2 + 3i}$$
$$= \frac{6 - 3 + 11i}{13}$$
$$= \frac{3}{13} + \frac{11}{13} i$$

Example

$$3 - i + i(2 + i) + \frac{6 + i}{3 + 4i} = 3 - i + 2i - 1 + \frac{(6 + i)(3 - 4i)}{25}$$
$$= 2 + i + \frac{22 - 21i}{25}$$
$$= \frac{72}{25} + \frac{4}{25} i$$

Example. Show that $-1 - \sqrt{5}i$ is a root of $x^3 + x^2 + 4x - 6 = 0$. Substituting $-1 - \sqrt{5}i$ into the left side of the equation, we have

$$
\begin{aligned}
(-1 - \sqrt{5}i)^3 &+ (-1 - \sqrt{5}i)^2 + 4(-1 - \sqrt{5}i) - 6 \\
&= (-1 - \sqrt{5}i)^2(-1 - \sqrt{5}i) + 1 + 2\sqrt{5}i - 5 - 4 - 4\sqrt{5}i - 6 \\
&= (-4 + 2\sqrt{5}i)(-1 - \sqrt{5}i) + 1 + 2\sqrt{5}i - 5 - 4 - 4\sqrt{5}i - 6 \\
&= 4 + 4\sqrt{5}i - 2\sqrt{5}i + 10 + 1 + 2\sqrt{5}i - 5 - 4 - 4\sqrt{5}i - 6 \\
&= 0
\end{aligned}
$$

The simplified version of Fortran we are presenting in this book does now allow its variables to take on complex numbers as values. An error message will result, for instance, if the computer is asked to evaluate SQRT(A) if A happens to be negative. Clearly, the computer does not recognize the symbol I as the imaginary unit. We can, however, compute the real and imaginary parts of complex numbers since they are real numbers, but we shall have to manually insert the i after the computer has computed them. We shall investigate this further in program examples that appear in subsequent sections of this chapter.

6.3 QUADRATIC FUNCTIONS AND EQUATIONS

A polynomial equation of the form $ax^2 + bx + c = 0$ is called a *quadratic* equation. There are several ways to solve for the two roots of such equations. One method is by factoring the left side into the form $(x - r_1)(x - r_2)$. We shall then have $(x - r_1)(x - r_2) = 0$. Now the product of numbers can be zero only if at least one of them is zero. Thus, it follows that either $x - r_1 = 0$ and hence that $x = r_1$ or that $x - r_2 = 0$ and hence that $x = r_2$. Therefore, r_1 and r_2 are the roots.

Example. Solve, by factoring, $3x^2 - 19x + 20 = 0$. Factoring, we have the following equation:

$(3x - 4)(x - 5) = 0$

Hence, $x = \frac{4}{3}$ or $x = 5$. The method of factoring is useful if the factors can easily be determined. But consider $2x^2 - 7x + 13 = 0$. The left side of this equation is evidently not easily factored.

We now present a method called the *quadratic formula* which can be used to solve any quadratic equation, whether or not it can be easily solved by the method of factoring. Consider the general equation $ax^2 + bx + c = 0$, with $a \neq 0$. Dividing through by a, we have $x^2 + bx/a + c/a = 0$. We move

c/a to the right side of the equation: $x^2 + bx/a = -c/a$. Next we *complete the square* on the left side by adding the quantity $b^2/4a^2$ to *both* sides of the equation. Now we have

$$x^2 + \frac{bx}{a} + \frac{b^2}{4a^2} = \frac{b^2}{4a^2} - \frac{c}{a}$$

$$= \frac{b^2}{4a^2} - \frac{4ac}{4a^2}$$

$$= \frac{b^2 - 4ac}{4a^2}$$

Factoring the left side gives

$$\left(x + \frac{b}{2a}\right)^2 = \frac{b^2 - 4ac}{4a^2}$$

Then taking the square root of both sides, we have

$$x + \frac{b}{2a} = \pm \frac{\sqrt{b^2 - 4ac}}{2a}$$

Thus

$$x = \frac{-b \pm \sqrt{b^2 - 4ac}}{2a}$$

the *quadratic formula*. To use the formula, we have merely to identify a, b, and c from the original equation and substitute them into the formula.

Example. Use the quadratic formula to solve $2x^2 - 7x + 13 = 0$. Here $a = 2$, $b = -7$, and $c = 13$. Thus, $x = (7 \pm \sqrt{49 - 104})/4 = (7 \pm \sqrt{-55})/4 = (7 \pm i\sqrt{55})/4 = 7/4 \pm i\sqrt{55}/4$. The roots are $7/4 + i\sqrt{55}/4$ and $7/4 - i\sqrt{55}/4$.
Note that the factored form of this equation is

$$(x - r_1)(x - r_2) = \left(x - \frac{7}{4} - i\frac{\sqrt{55}}{4}\right)\left(x - \frac{7}{4} + i\frac{\sqrt{55}}{4}\right) = 0$$

No wonder we could not *easily* solve this equation by factoring.

Example. Solve $x^2 + 2x - 15 = 0$. Here $a = 1$, $b = 2$, $c = -15$. So,

$$x = \frac{-2 \pm \sqrt{4 + 60}}{2} = \frac{-2 \pm 8}{2}$$

Hence, $x = (-2 + 8)/2 = 3$ or $x = (-2 - 8)/2 = -5$. The roots are 3 and -5.

The quantity $b^2 - 4ac$ that occurs in the quadratic formula is called the *discriminant* of the equation. Since $\sqrt{b^2 - 4ac}$ occurs in the formula, it is clear that the roots r_1 and r_2 will be real and unequal if $b^2 - 4ac > 0$ and will be complex if $b^2 - 4ac < 0$. In the case $b^2 - 4ac = 0$, the roots are real and *double*.

Example. Solve $x^2 - 6x + 9$ using the quadratic formula. $a = 1$, $b = -6$, and $c = 9$. Hence, $x = (6 \pm \sqrt{36 - 36})/2 = (6 \pm 0)/2$. The roots are then $(6 + 0)/2 = 3$ and $(6 - 0)/2 = 3$. Clearly, 3 is a double root. The equation has two equal roots.

Let $f(x)$ be any function of x. Then the function $a[f(x)]^2 + b[f(x)] + c$ is in quadratic form. It is clearly just the quadratic function $ax^2 + bx + c$ with x replaced by $f(x)$. The *intermediate* solution of the equation $a[f(x)]^2 + b[f(x)] + c = 0$ is just the quadratic formula with x replaced by $f(x)$. That is, $f(x) = (-b \pm \sqrt{b^2 - 4ac})/2a$. This is, however, another equation which must be further *solved for x*.

Consider the equation $4(x - 5)^2 + 3(x - 5) - 8 = 0$. Letting $f(x) = x - 5$, we have $x - 5 = (-3 \pm \sqrt{9 + 128})/8 = -\frac{3}{8} \pm \sqrt{137}/8$. So, $x = 5 - \frac{3}{8} \pm \sqrt{137}/8 = \frac{37}{8} \pm \sqrt{137}/8$. That is, $x = (37 + \sqrt{137})/8$ or $x = (37 - \sqrt{137})/8$.

Example. Solve $1/x^4 + 4/x^2 - 21 = 0$. We let $f(x)$ be $1/x^2$, and the equation is in quadratic form. Then $1/x^2 = (-4 \pm \sqrt{16 + 84})/2 = (-4 \pm 10)/2$. Thus, we have either $1/x^2 = 3$ or $1/x^2 = -7$. In the first case, $x^2 = \frac{1}{3}$ and $x = \pm \sqrt{3}/3$. In the second case, $x^2 = -\frac{1}{7}$ and $x = \pm(\sqrt{7}/7)i$.

Example. Solve $15/x - 9\sqrt{3/x} + 4 = 0$. This equation can be written $5(\sqrt{3/x})^2 - 9(\sqrt{3/x}) + 4$ and hence is in quadratic form with $f(x) = \sqrt{3/x}$. Using the quadratic formula, we have $\sqrt{3/x} = (9 \pm \sqrt{81 - 80})/10 = (9 \pm 1)/10$. Therefore, $\sqrt{3/x} = 1$ or $\sqrt{3/x} = \frac{4}{5}$. In the first case, $3/x = 1$ and $x = 3$. In the second case, $3/x = \frac{16}{25}$ and $x = \frac{75}{16}$. You can check that each of these is a root by substituting it into the original equation.

6.4 ANALYTIC GEOMETRY OF QUADRATIC FUNCTIONS; CURVEFITTING

The graph of the quadratic function $y = ax^2 + bx + c$, $a \neq 0$, is called a *parabola*. Consider, for example, $y = x^2 + 3$. Its graph can be sketched by first finding an appropriate number of points that are on it. The following table of values can be used:

x	−3	−2	−1	0	1	2	3
y	12	7	4	3	4	7	12

The graph appears in Fig. 6.1.

The low point (or high point) of these parabolas is called the *vertex* of the parabola. The point $(0,3)$ is the vertex of the parabola of Fig. 6.1. The curve is clearly symmetric with respect to the y axis.

The equation of any parabola of this type can be written in the *standard form*

$$y - y_0 = p(x - x_0)^2$$

When the equation is written in this form, the point with coordinates (x_0, y_0) is the vertex. If p is greater than zero, the parabola opens upward; if p is negative, the parabola opens downward.

Example. Suppose we have $y = -4x - x^2$. We place this equation in standard form in the following way: $y = -(x^2 + 4x) = -(x^2 + 4x + 4) + 4 = -(x + 2)^2 + 4$. Thus, $y - 4 = -(x + 2)^2$. From this equation we read that the vertex is at $(-2,4)$ and since $p = -1$, the parabola opens downward. This information, together with a few more points which lie on the parabola, will enable us to draw a perfectly suitable sketch. Often the points where the curve

Fig. 6.1

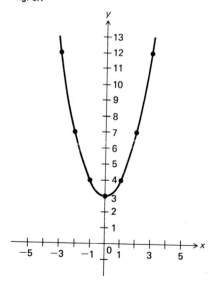

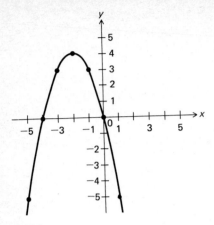

Fig. 6.2

crosses the coordinate axes are useful. Clearly, the parabola above crosses the x axis at the point where $y = 0$. Setting $y = 0$ in the equation, we solve for x and discover that the points $(0,0)$ and $(-4,0)$ are the x intercept points. Similarly, setting $x = 0$ in the equation, we discover that the curve crosses the y axis at $(0,0)$. The sketch of the graph of $y - 4 = -(x + 2)^2$ appears in Fig. 6.2. We note that the line $x = -2$ is now the axis of symmetry of the parabola.

Example. Find the vertex and intercepts of the parabola $y = 4x^2 - 8x + 5$ and sketch it.

We place the equation in standard form as follows: $y = 4(x^2 - 2x) + 5 = 4(x^2 - 2x + 1) + 5 - 4 = 4(x - 1)^2 + 1$. Thus, $y - 1 = 4(x - 1)^2$, the vertex is at $(1,1)$, and the parabola opens upward, since $p = 4 > 0$. As for the intercepts, when $x = 0$, $y = 5$, and so the point $(0,5)$ is the y intercept point. But now when $y = 0$, we have $4x^2 - 8x + 5 = 0$, so that $x = (8 \pm 4i)/2$, using the quadratic formula. This nonreal value for x is algebraic evidence that the parabola does not intersect the x axis. The sketch in Fig. 6.3 substantiates this evidence geometrically.

The quadratic function $ax^2 + bx + c$ is often used for curvefitting purposes. Recall the analysis we gave when we discussed the fitting of data to a straight line by the method of least squares. If the data, when plotted, look as if a parabola would fit them better than a line would, the method of least squares will once again provide the means for this fitting. As we shall soon see, the arithmetic calculations involved are straightforward but tend to be somewhat laborious. The use of the computer is thus indicated in order to

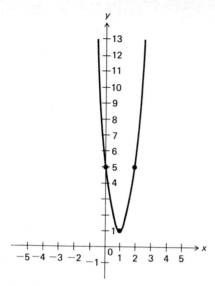

Fig. 6.3

relieve us of some of the work. However, we plunge forward and solve the problem by hand calculations.

Consider the (empirical) data

x	−2	−1	0	1	2	3
y	3	1	0	1	1	3

The points are plotted in Fig. 6.4, and a reasonable parabola has been *sketched* in just to see whether the idea is feasible. It evidently is. The technique we now describe will, of course, determine the unique parabola which is "best fitting" by the method of least squares and will undoubtedly not be at all the parabola we have just sketched in.

Just a note about the use to which one would be able to put the equation of the best-fitting parabola $y = ax^2 + bx + c$. It can be used to *interpolate* values of x, that is, to *approximate* the value y would have for values of x in the domain $[-2,3]$ which are not data values.

Suppose the parabola we seek has equation $y = ax^2 + bx + c$. The coefficients a, b, and c of the best-fitting parabola are found by solving the following set of simultaneous linear equations for a, b, and c.

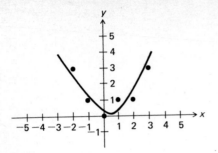

Fig. 6.4

$$\begin{cases} \left[\sum_{i=1}^{n} x_i^2\right] a + \left[\sum_{i=1}^{n} x_i\right] b + nc = \left[\sum_{i=1}^{n} y_i\right] \\ \left[\sum_{i=1}^{n} x_i^3\right] a + \left[\sum_{i=1}^{n} x_i^2\right] b + \left[\sum_{i=1}^{n} x_i\right] c = \left[\sum_{i=1}^{n} x_i y_i\right] \\ \left[\sum_{i=1}^{n} x_i^4\right] a + \left[\sum_{i=1}^{n} x_i^3\right] b + \left[\sum_{i=1}^{n} x_i^2\right] c = \left[\sum_{i=1}^{n} x_i^2 y_i\right] \end{cases}$$

Here the (x_i, y_i) are the given data points, and n is the number of data points. The various summations that occur must be computed from the data, and then the three simultaneous linear equations resulting must be solved for a, b, and c. To facilitate finding all these sums, the following tabular method is displayed for the data just given $(n = 6)$:

x	y	x²	x³	x⁴	xy	x²y
−2	3	4	−8	16	−6	12
−1	1	1	−1	1	−1	1
0	0	0	0	0	0	0
1	1	1	1	1	1	1
2	1	4	8	16	2	4
3	3	9	27	81	9	27
3	9	19	27	115	5	45

The entries in the last row are the various sums that are requested as coefficients in the equations above. We then have the following set of equations to solve:

$$\begin{cases} 19a + 3b + 6c = 9 \\ 27a + 19b + 3c = 5 \\ 115a + 27b + 19c = 45 \end{cases}$$

As noted above, here is where some sort of automatic calculator would be helpful. But since the numbers involved are not too unwieldy, we proceed by hand. We use the simple method of elimination. Multiplying the second equation through by 2 and subtracting from the first equation, we have $35a + 35b = 1$. We again eliminate c by multiplying the first equation by 19, the third equation through by 6, and subtracting. We then have $329a + 105b = 99$. We must then solve the set

$$\begin{cases} 329a + 105b = 99 \\ 35a + 35b = 1 \end{cases}$$

Multiplying the second equation through by 3 and subtracting from the first equation, we easily eliminate b. We then have $224a = 96$, that is, $a = 3/7$. Then substituting into $a + b = 1/35$, we have $b = 1/35 - 3/7 = -2/5$. Now substituting both a and b into the first original equation, $19a + 3b + 6c = 9$, we have $57/7 - 6/5 + 6c = 9$, from which $c = 12/35$.

The best-fitting parabola has the equation $y = 3x^2/7 - 2x/5 + 12/35$. More analysis of this example is requested in the exercise set that follows.

EXERCISES

6.1 If $z_1 = 3 - 4i$, $z_2 = -1 - 2i$, $z_3 = 2i$, evaluate the following expressions and give the result in the form $a + bi$:

(a) $z_1 + 3z_2$

(b) $2z_1z_2 - 4z_1{}^2$

(c) $z_3{}^5$

(d) $\dfrac{z_1 + 3i}{z_2 + 2}$

(e) $\dfrac{z_1 + z_2}{z_1 + z_3}$

(f) $4iz_1 - 5iz_2 + 6iz_3$

(g) $(z_3/2)^{47}$

6.2 Solve the following quadratic equations using factoring:

(a) $x^2 - 4x + 3 = 0$

(b) $2t^2 - t = 21$

(c) $10x^2 - 11x - 6 = 0$

(d) $4s^2 - 20s + 21 = 0$

(e) $9y^2 - 52y + 35 = 0$

6.3 Solve the following quadratic equations using the quadratic formula:

(a) $x^2 - 8x + 40 = 0$
(b) $x^2 + 3x - 10 = 0$
(c) $4t^2 - 4t + 1 = 0$
(d) $2z^2 - 4z = 7$
(e) $3x^2 + 7x + 12 = 0$
(f) $7x^2 - 4x + 3 = 0$
(g) $x^2 + 4x + 6 = 0$
(h) $y^2 = -7y$
(i) $4q^2 + 9 = 0$
(j) $5r^2 - 5r + 7 = 0$

6.4 Solve the following equations which are in quadratic form:

(a) $\dfrac{1}{x^2} + \dfrac{4}{x} + 3 = 0$

(b) $x - 2\sqrt{x} - 8 = 0$

(c) $\dfrac{48}{(3-x)^2} + \dfrac{20}{3-x} - 2 = 0$

(d) $t^4 - 3t^2 - 10 = 0$

(e) $x^{4/3} - 10x^{2/3} + 9 = 0$

(f) $3\left(\dfrac{x-1}{x+2}\right)^2 - 25\,\dfrac{x-1}{x+2} = -28$

6.5 Place the equation of each of the following parabolas in standard form, find the coordinates of the vertex, determine whether it opens upward or downward, find its x and y intercepts, and by finding enough additional points, sketch the parabola:

(a) $y = 7x^2 - 28x + 25$
(b) $y = -2x^2 - 4x - 1$
(c) $128y = 63 + 8x - 16x^2$
(d) $y = 0.6x - x^2 - 0.29$
(e) $y = 8x - 2x^2 - 1$
(f) $y = 5x^2 + 4x - 7$

6.6 Refer to the parabola $y = 3x^2/7 - 2x/5 + {}^{12}/_{35}$ which was the best-fitting parabola for the curvefitting example given in the text. Place this equation in standard form and thus determine its vertex. Then sketch the parabola on a graph which also contains the original data points.

6.7 To check just how well the parabola of Exercise 6.6 fits the data set given in the text, calculate the value of y on the parabola for each x data value and compare with the given y data.

6.8 Use the parabola of Exercise 6.6 to estimate y when $x = \frac{3}{2}$ and when $x = -\frac{1}{2}$.

6.9 Use the method of least squares to fit each of the following data sets to the parabola $y = ax^2 + bx + c$:

(a)

x	y
4	0
6	1
7	3

(b)

x	y
1	3
2	0
3	2
4	4

(c)

x	y
-1	-1
0	0
1	1
2	3
3	8

6.10 Write a complete Fortran program to solve *any* quadratic equation $ax^2 + bx + c = 0$, $a \neq 0$. Read the values of a, b, and c on one card, 3F10.0. Print the real and imaginary parts of each of the roots. Allow any number of cards containing a, b, and c to be consecutively read so that the program can solve any number of quadratic equations.

6.5 POLYNOMIAL EQUATIONS OF DEGREE HIGHER THAN 2

Polynomial equations of degree higher than 2 cannot in general be easily solved. The cubic equation (degree 3) can be solved exactly using a formula of much more difficult form than the quadratic formula, but this cubic formula is not now currently in vogue. Instead, numerical methods to find the real roots of such equations are used, methods which can apply in general to polynomial equations of any degree. We shall present an introduction to some of these methods in this section, but we must leave quite a lot of material to await your eventual study of a full-blown course in numerical methods.

Most of the numerical methods for finding a real root of a polynomial equation start out with the knowledge of the approximate location of the root. Then the method proceeds to zero in more and more closely on the root in order to find it to a good degree of accuracy. A simple method of locating a real root is now described. Let $f(x)$ be a polynomial function and suppose $x_2 > x_1$. If $f(x_2) > 0$ and $f(x_1) < 0$, then the equation $f(x) = 0$ has a real root r such that $x_1 < r < x_2$. The same result follows if $f(x_2) < 0$ while $f(x_1) > 0$.

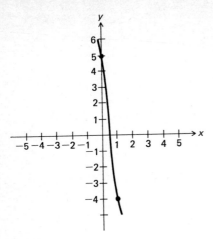

Fig. 6.5

For example, suppose $f(x) = x^3 - 3x^2 - 7x + 5$. Clearly, $f(0) = 5$ while $f(1) = -4$. Therefore, $f(x) = 0$ has at least one real root between $x = 0$ and $x = 1$. The graph of Fig. 6.5 shows the geometric idea behind this rule. On the graph the points $(0, f(0))$ and $(1, f(1))$ are plotted, and a smooth single-valued curve has been sketched between them to represent approximately the graph of $f(x)$ between 0 and 1. Clearly, $f(x)$ must cross the x axis somewhere between 0 and 1, or else how could it get from above the axis to below the axis? At the value of x where the graph crosses the x axis, $y = 0$; that is, $f(x)$ equals 0, and this value of x is the desired real root. We, of course, have only a reasonably good approximate idea of what the root is, but at least we now have a starting place for a refining process to find the root more precisely. We note that the root is approximately halfway between 0 and 1. Let us check $x = \frac{1}{2}$ as the approximate value of the root. We have $f(\frac{1}{2}) = (\frac{1}{2})^3 - 3(\frac{1}{2})^2 - 7(\frac{1}{2}) + 5 = \frac{7}{8}$. So, $\frac{1}{2}$ is not the root but is close to the root. Now, since $f(1) = -4$ and $f(\frac{1}{2}) = \frac{7}{8}$, we see that the root is between $\frac{1}{2}$ and 1 and is evidently much closer to $\frac{1}{2}$ than it is to 1.

Having located the desired real root at least between two integers, we can choose from among many techniques of numerical analysis to squeeze down on the root. Many of these make use of the differential calculus, but there are methods that use geometrically oriented notions not so involved. One method that will always work is called the method of *bisection*. It depends only on a simple geometric construction which we now explain.

Suppose we have isolated the real root between two integers x_1 and x_2. Suppose $f(x_1) > 0$ and $f(x_2) < 0$ and the situation is shown in Fig. 6.6. The root is r. Suppose we now *bisect* the *interval* from x_1 to x_2; that is, let x_3 be $(x_1 + x_2)/2$. We note that x_3 is shown on the figure. We now calculate $f(x_3)$.

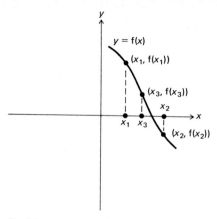

Fig. 6.6

If it is 0, then x_3 is the root and our work is done. But if $f(x_3) > 0$, the root must be between x_3 and x_2 since $f(x_2) < 0$. Similarly, if it had turned out that $f(x_3)$ had been negative, the root r would be between x_1 and x_3 [since $f(x_1)$ is positive]. In any case we have somewhat closed in on the root. If we carry on, following the graph of Fig. 6.6 as our example, we would then proceed to repeat the process. Again, bisect the interval in which we know the root lies, this time the interval between x_3 and x_2. Suppose we let $x_4 = (x_2 + x_3)/2$. Again we calculate $f(x_4)$ and by observing its sign decide whether the root now lies between x_3 and x_4 or between x_4 and x_2. The process consists of repeated bisection operations as often as desired. Clearly, we shall close down on the root, and even though the process may be slow and laborious, it cannot fail to converge to the root. As usual, when laborious but well-defined calculations confront us, we immediately think of making use of a computer to do the work. However, let us present just a few steps of the method of bisection as it is used to solve the cubic equation $x^3 - 3x^2 - 7x + 5 = 0$ originally given for the root between 0 and 1.

Let us start with $x_1 = \frac{1}{2}, x_2 = 1$. Recall that $f(\frac{1}{2}) = 0.875$ and $f(1) = -4$. Now $x_3 = \frac{1}{2}(x_1 + x_2) = 0.75$. We calculate $f(0.75)$ as -1.52. Thus the root must be between 0.5 and 0.75. Again, we calculate $x_4 = \frac{1}{2}(0.5 + 0.75) = 0.625$. Now $f(0.625) = -0.30$ so that the root must be between 0.5 and 0.625, since $f(0.5) = 0.875$. Perhaps you will care to proceed and show that the root is approximately 0.599.

A very useful iterative procedure, which is often used since it converges to the root rather rapidly, is called *Newton's method*. The formula we are about to give is based on the differential calculus and will not be derived here. It will be sufficient to note first that associated with each polynomial function $f(x) = a_n x^n + a_{n-1} x^{n-1} + \cdots + a_2 x^2 + a_1 x + a_0$ is another polynomial function called its *derivative* and denoted by $f'(x)$ and $f'(x) = na_n x^{n-1} + (n-1)a_{n-1}x^{n-2} + \cdots + 2a_2 x + a_1$.

Example. Given $f(x) = 8x^4 + 3x^2 - 7x + 9$, then $f'(x) = 32x^3 + 6x - 7$. It is found by simply multiplying the coefficient of each term by the corresponding power of x and then reducing the power by 1.

Now suppose that $f(x) = 0$ has a real root near x_1. This fact is to be determined earlier as we did in the previous method. Then, Newton's method consists of evaluating the iterative formula

$$x_{i+1} = x_i - \frac{f(x_i)}{f'(x_i)} \qquad i = 1, 2, 3, 4, \ldots$$

The sequence $x_2, x_3, x_4, \ldots$ thus generated presumably converges to the root desired. One naturally takes x_1 as the first approximate value.

Let us take an example for which we can find the root by analytic methods just so we can check the result. Let $f(x) = x^2 - 3x + 1 = 0$. Since $f(2) = -1$ and $f(3) = 1$, this equation has a real root between $x = 2$ and $x = 3$. Let us take $x_1 = 3$ and apply Newton's method. Now, $f(x) = x^2 - 3x + 1$ so that $f'(x) = 2x - 3$. Newton's formula for this problem is

$$x_{i+1} = x_i - \frac{x_i^2 - 3x_i + 1}{2x_i - 3}$$

$$x_1 = 3$$

So, letting $i = 1$, we have

$$x_2 = x_1 - \frac{x_1^2 - 3x_1 + 1}{2x_1 - 3}$$

$$= 3 - \frac{9 - 9 + 1}{6 - 3}$$

$$= \frac{8}{3}$$

Then, $i = 2$ and

$$x_3 = x_2 - \frac{x_2^2 - 3x_1 + 1}{2x_2 - 3}$$

$$= \frac{8}{3} - \frac{64/9 - 8 + 1}{16/3 - 3}$$

$$= \frac{55}{21}$$

Finally, $^{55}/_{21} \doteq 2.619$.

We may continue in the same manner to increase the accuracy, but, of course, we must now turn to approximate decimal numbers resulting in some pretty hard calculations. For the fun of it, let us now solve the original equation using the quadratic formula. Obviously, the root we are seeking is exactly $(3 + \sqrt{5})/2$. Using tables, we have $\sqrt{5} \doteq 2.236$ so that the root is $\doteq (3 + 2.236)/2$. This is 2.618, showing that our result above is pretty good.

Other iterative methods for solving polynomial equations appear in the exercise set below.

EXERCISES

6.11 Use the method of bisection to solve the following:

(a) $2x^3 + 4x - 7 = 0$ for the real root between 1 and 2
(b) $x^4 - 3x^2 - 15x - 16$ for the root between 3 and 4
(c) $x^3 - 87 = 0$ for the root between $x = 4$ and $x = 5$

6.12 Solve the following equations using Newton's method:

(a) $x^3 - 3x - 15 = 0$ for its positive real root
(b) $x^4 - 8x^3 + 7x - 12 = 0$ for the root between -1 and -2
(c) $x^3 + 2x^2 - 7x + 5 = 0$ for any real root
(d) $x^4 - 3x^3 + 2x^2 + 4x - 7 = 0$ for the root between 1 and 2

6.13 Other iterative methods can be concocted depending upon the ingenuity of the concocter. For example, to solve $x^3 + 3x - 5 = 0$ for the root near 1, one could dream up the iterative formula

$$x_{i+1} = x_i - \frac{x_i^3 + 3x_i - 5}{10} \qquad x_1 = 1$$

See how this formula works by applying it several times.

6.14 To solve $x^4 + 3x^2 - 7 = 0$, one could maneuver the equation into

$$x^2(x^2 + 3) = 7 \qquad x^2 = \frac{7}{x^2 + 3} \qquad x = \left(\frac{7}{x^2 + 3}\right)^{1/2}$$

To find the root near 1, one uses this last formula in an iterative manner by writing it

$$x_{i+1} = \sqrt{\frac{7}{x_i^2 + 3}} \qquad x_1 = 1$$

Use this iterative formula several times to test its convergence. Solve the original equation exactly by recalling that it is in quadratic form and using the quadratic formula.

6.15 Write a complete Fortran program to solve the fourth-degree equation of Exercise 6.14 using the method given there.

6.16 Write a complete Fortran program to solve the cubic equation given in Exercise 6.13 using the iterative formula given there.

6.17 Solve any of the equations of Exercise 6.12 by writing a complete Fortran program.

7

Exponential and Logarithmic Functions

7.1 INTRODUCTION. DEFINITIONS

The function $y = a^x$, $a > 0$, $a \neq 1$, is called the *exponential function*. The domain of this function is the set of all real numbers, that is, $x \in (\infty, -\infty)$. Let us discover some properties of this function by drawing its graph. First, let $a > 1$. For example, $y = 2^x$. We easily calculate the following table of values:

x	y
-2	$2^{-2} = 1/4$
-1	$2^{-1} = 1/2$
0	$2^0 = 1$
1	$2^1 = 2$
2	$2^2 = 4$
3	$2^3 = 8$

We plot these points and draw the graph in Fig. 7.1. From the graph we note that y is positive for all x. y is always increasing with x; that is, the graph rises continuously as x moves from left to right. y increases rather rapidly as x becomes more positive and larger. Evidently $y = 1$ is the y intercept, and the graph has no x intercept. As x becomes more and more negative, y be-

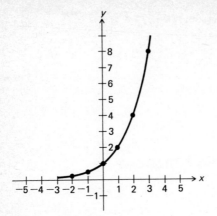

Fig. 7.1

comes closer and closer to zero. We say that a line like $y = 0$ is a *horizontal asymptote*.

Now, suppose $0 < a < 1$; for example, $y = (1/3)^x$. A table of values and the graph follow.

x	y
−2	$(1/3)^{-2} = 9$
−1	$(1/3)^{-1} = 3$
0	$(1/3)^0 = 1$
1	$(1/3)^1 = 1/3$
2	$(1/3)^2 = 1/9$

Fig. 7.2

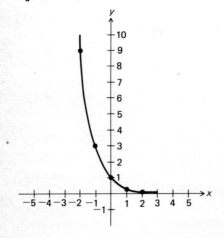

For our subsequent use, we recall the "laws of exponents."

$$a^x a^y = a^{x+y}$$

$$\frac{a^x}{a^y} = a^{x-y}$$

$$(a^x)^y = a^{xy}$$

Now we begin our development of the logarithmic function. We write $x = \log_a y$ if $a^x = y$ and read this first equation "x is the logarithm to the base a of y." It appears that x is the logarithm of y if the base a raised to the power x is y. The relations $x = \log_a y$ and $a^x = y$ are identical and have the *same* graphs.

Suppose we algebraically interchange x and y in $x = \log_a y$ to get $y = \log_a x$. We then have normal variable names, and y is the logarithmic function of x. That is, now y is the dependent variable, and x is the independent variable. The graph of $y = \log_a x$ can be determined by noting the geometric significance of the algebraic interchange of variables. The graph of $y = \log_a x$ is simply the graph of $y = a^x$ *reflected into the line* $y = x$. The reason for this is that (y,x) satisfies $y = \log_a x$ whenever (x,y) satisfies $y = a^x$. Figure 7.3 shows this geometric juxtaposition.

From the graph of $y = \log_a x$ we note that the function $y = \log_a x$ is not defined if x is 0 or negative. The logarithm of 1 is zero since the x intercept of the graph is $x = 1$. The logarithm of numbers less than 1 is negative and

Fig. 7.3

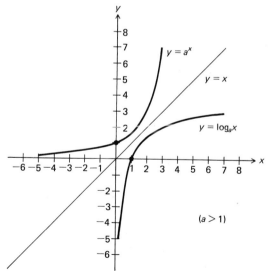

for numbers more than 1 is positive. Recall that the base a is a positive real number.

To make clear the equivalence of $y = \log_a x$ and $x = a^y$, suppose we want to calculate the value of $y = \log_5 25$. We then have $5^y = 25$, and inspection reveals that $y = 2$. Thus, y is the power to which 5 must be raised to give 25. Again, consider $y = \log_{1/2} 8$. Thus, $(1/2)^y = 8$, $2^y = 1/8$, and $y = -3$. We can also ask, "If $3 = \log_a 8$, what is a?" Since we know that $a^3 = 8$, it is clear that the base $a = 2$.

EXERCISES

7.1 Find y if

(a) $y = \log_3 27$
(b) $y = \log_7 (1/49)$
(c) $y = \log_{10} 0.001$
(d) $y = \log_{1/4} 16$
(e) $y = \log_9 3$
(f) $y = \log_5 0.008$

7.2 Find the base a if

(a) $\log_a 8 = 3$
(b) $\log_a 8 = -3$
(c) $\log_a 27 = 3/2$

These examples and exercises have used special numbers so that the result would be easy to find by inspection. Suppose one were to be asked, "What is $\log_3 7$?" In other words, to what power must 3 be raised to give 7? This power is an irrational number, and the best we can do is to find a rational approximation to it. Possibly this rational answer appears in some kind of a *table* of logarithms. You have already seen an entirely similar situation occur in trigonometry. Although the sine of 30° and the sine of 45° are easily found without using trigonometric tables, it is not easy to approximate the sine of 17° without a table (or at least a slide rule). So, as we suspect, tables of the values of the logarithmic functions have been constructed, and from them we can read approximate values of both the logarithmic and exponential functions. We observe that since there are infinitely many bases we could use, probably we should restrict the bases used in tables to just one or two special ones. This is precisely what has been done, the bases chosen being 10 and e. The latter is a very special irrational number which has been given a special name, as has π. Logarithms using base 10 are called *common* logarithms, and those using base e are called *natural* logarithms.

If we are restricted numerically to two bases, then how do we calculate logarithms which use bases other than these? Certain identities which will now be derived permit this. Let $p = \log_a N$ and $q = \log_b N$. Then $a^p = N$ and $b^q = N$. Thus, $a^p = b^q$ or $a = b^{q/p}$. Therefore, q/p is $\log_b a$. But q/p is also $\log_b N / \log_a N$. We finally have the identity:

$$\log_b N = (\log_a N)(\log_b a)$$

In the light of our previous discussion, if we let $a = 10$, then

$$\log_b N = (\log_{10} N)(\log_b 10)$$

We can find the logarithm of N to any base b if we know its common logarithm *and* the logarithm of 10 to the base b.

Since most conversions of bases are between bases 10 and e, it would be sufficient to know that $\log_e 10 \doteq 2.30259$ and $\log_{10} e \doteq 0.43429$.

Example. Suppose we know that $\log_{10} 43 = 1.63347$. Then $\log_e 43 = \log_{10} 43 \cdot \log_e 10 = 1.63347(2.30259) = 3.76120$. This result can be verified by consulting a table of natural logarithms.

7.2 CALCULATIONS INVOLVING LOGARITHMIC AND EXPONENTIAL FUNCTIONS

We first derive three basic properties of the logarithmic function. Let $x = \log_a A$ and $y = \log_a B$. Then $a^x = A$ and $a^y = B$. Then $AB = a^x a^y = a^{x+y}$. Taking logarithms base a of both sides, we have $\log_a AB = x + y = \log_a A + \log_a B$, or

$$\log_a AB = \log_a A + \log_a B \tag{7.1}$$

Similarly, $a^x/a^y = a^{x-y}$. Therefore,

$$\log_a \frac{A}{B} = \log_a A - \log_a B \tag{7.2}$$

Since $A^r = (a^x)^r = a^{rx}$, we also have

$$\log_a A^r = r \log_a A \tag{7.3}$$

In essence, Eq. (7.1) reduces the problem of multiplying two real numbers to that of adding their logarithms and then finding the real number which has this sum for its logarithm. Equation (7.2) reduces the problem of division to that of subtraction, and Eq. (7.3) reduces the problem of raising to a power to multiplying a logarithm by the power. In each case, of course, there is the additional problem of converting the logarithmic result back to the number which has this logarithm.

We shall presume that you have had some experience with the use of logarithms as tools of calculation and that you know that the design of the slide rule is based on logarithmic properties. Here we are concerned more with the logarithmic and exponential *functions*.

The natural logarithm $\log_e x$ and exponential e^x use base e. It is conventional to write $\log_e x$ as $\ln x$ (with the base now assumed to be e). Incidentally, advanced methods reveal that $e \doteq 2.718281828$. Interestingly enough, the natural logarithm lives up to its description "natural." It is the most used type of logarithm in applied mathematics and engineering. Your differential and integral calculus classes will reveal its importance and why it has been given the adjective "natural." Many physical quantities obey physical laws. You will find, for example, that the current in a certain circuit decays exponentially, that is, $i = i_0 e^{-kt}$, and that a certain population increases exponentially, that is, $p = p_0 e^{kt}$. The atmospheric pressure P in pounds per square inch at an elevation of h miles above sea level varies approximately according to $p = 14.7 e^{-kh}$. One may solve this last equation for h. Taking the natural logarithm of both sides, we have $\ln p = \ln 14.7 e^{-kh} = \ln 14.7 + \ln e^{-kh} = \ln 14.7 - kh$. Thus, $h = (\ln 14.7 - \ln p)/k$. If one knew k, one could use this last relation to find the elevation for a given pressure p.

EXERCISES

7.3 Suppose $p = 14.7 e^{-0.180h}$. Use natural logarithm tables to find the elevation h (in miles) at which p is 7.35 lb/in.^{2}.

7.4 Suppose $i = 5.0 e^{-3.8t}$. Use exponential tables to find i when $t = 1.5$. i is in amperes, t in seconds.

7.3 TRANSCENDENTAL EQUATIONS INVOLVING LOGARITHMIC AND EXPONENTIAL FUNCTIONS

Recall that a reasonable method for solving certain types of equations is the iterative process in which we write the equation in the form $x = g(x)$, find a good starting approximation for the root we desire, call it x_1, and iterate using the formula

$$x_{i+1} = g(x_i) \qquad i = 1, 2, 3, \ldots$$

Not every iterative formula will converge, however.

Consider the equation $e^x + 3x = 20$. Investigation in exponential tables and a few rapid calculations reveal that the equation has a root near 2.5. We shall choose this for x_1. Now we rewrite the equation in the form $x = \ln(20 - 3x)$ and set up the iterative formula

$$x_{i+1} = \ln(20 - 3x_i) \qquad i = 1, 2, 3, 4, \ldots$$

Calculation reveals that $x_2 = 2.5257$, $x_3 = 2.5180$, and so the process seems to be converging nicely toward the root.

Let us write a Fortran program to solve this problem. Recall that Fortran has built-in *library* functions that automatically calculate the values of functions like e^x, ln x, sin x, etc. We write EXP(X) for e^x and ALOG(X) for ln x. In each of these, X must be a real constant or a real or mixed expression. For example, the Fortran statement Y = EXP(−3.*X) tells the computer to compute the value of e^{-3x} for the current value of x and store the result at Y.

Since we have decided that this particular formula converges when $x_1 = 2.5$, let us write the program to iterate 20 times and then print the value of the then current value of x. The program could be as follows:

```
    X = 2.5
    I = 0
5   X = ALOG(20. − 3.*X)
    I = I + 1
    IF (I − 20) 5, 5, 6
6   WRITE (3, 1) X
1   FORMAT (1X, E13.6)
    END
```

First, x is initialized at 2.5. The counter I is set at 0. Note the statement numbered 5. The righthand side is evaluated using $x = 2.5$. Then the result is stored at X, erasing the original value of X and replacing it with the new approximation. The counter is advanced by 1, and since I is still considerably less than 20, transfer is made to the statement numbered 5. The new approximation is used to calculate the righthand side once again. This next approximation again replaces the old, and the process is repeated. By the time the output step is arrived at, only the very last approximate value remains and is called X. This is printed and the program ENDs.

EXERCISES

7.5 Write a complete Fortran program using the same iterative formula to solve the equation $e^x + 3x = 20$. Again, let $x_1 = 2.5$. This time, however, print x after each *five* iterations. Iterate a total of 20 times.

7.6 Investigate writing the equation $e^x + 3x = 20$ in the form $x = (20 - e^x)/3$. Set up the iterative formula $x_{i+1} = (20 - e^{x_i})/3$, $i = 1, 2, 3, \ldots$. Using exponential tables, start calculating by hand starting with $x_1 = 2.5$. Does the process seem to be converging? Write a complete Fortran program to do the calculating and print the result of each iteration, that is, each x_i, for $i = 2, 3, 4, \ldots, 20$.

7.7 The equation $\log_{10} 3x + x^2 = 10$ has a real root near $x = 3$.

(a) Show that this equation can be rewritten in the form

$$x = \sqrt{10 - \log_{10} 3x}$$

or in the form $x = \dfrac{1}{3}(10^{10-x^2})$.

(b) Write a complete Fortran program using the formula

$$x_{i+1} = \frac{1}{3}(10^{10-x_i^2})$$

$x_1 = 3$, to solve the equation. Iterate 20 times and print the result of each iteration.

(c) Write a complete Fortran program using the formula $x_{i+1} = (10 - \log_{10} 3x_i)^{1/2}$, $x_1 = 3$, to solve the equation. Note that the computer will compute ALOG(X) as the *natural* logarithm of x. Recall from Sec. 7.1 how to change from base e to base 10. In any case, iterate 20 times and print the result of each iteration.

Recall Newton's method for solving the equation $f(x) = 0$ for the real root near x_1:

$$x_{i+1} = x_i - \frac{f(x_i)}{f'(x_i)} \qquad i = 1, 2, 3, \ldots$$

If $f(x)$ has a term be^{ax}, then the corresponding term in $f'(x)$ is bae^{ax}. If $f(x)$ has a term $b \ln(ax)$, then $f'(x)$ contains the term b/x. Recall that $f'(x)$ contains the term anx^{n-1} for each term ax^n contained in $f(x)$. For example, suppose $f(x) = 4e^{2x} - 5x^3 + 3 \ln x$. Then $f'(x) = 8e^{2x} - 15x^2 + 3/x$. The equation $4e^{2x} - 5x^3 + 3 \ln x = 0$ has a real root near $x = 1$. Newton's iterative formula is, therefore,

$$x_{i+1} = x_i - \frac{4e^{2x_i} + 3 \ln x_i - 5x_i^3}{8e^{2x_i} + 3/x_i - 15x_i^2} \qquad x_1 = 1$$

```
    I = 1
    X = 1.0
1   U = 4.*EXP(2.*X) + 3.*ALOG(X) − 5.*X**3
    V = 8.*EXP(2.*X) + 3./X − 15.*X**2
    X = X − U/V
    I = I + 1
    IF (I − 10) 1, 1, 2
2   WRITE (3, 3) X
3   FORMAT (1X, E13.6)
    END
```

7.8 Write the equation of Exercise 7.5 as $e^x + 3x - 20 = 0$. Write a complete Fortran program using Newton's iterative method to solve for the root near 2.5. Iterate 10 times.

7.4 LOG AND SEMILOG GRAPH PAPER. CURVEFITTING

Log and semilog graph paper is often used in applications where the graphing of exponential or logarithmic functions is involved. To understand its construction and use, let us start with a *logarithmic* coordinate scale instead of the ordinary *uniform* scale we have used up to now. As we know, the uniform scale is one in which each unit length represents the same change in the valu? of the variable. A *logarithmic* scale looks like this:

On this scale if we mark $x = 4$, we are actually plotting the logarithm of 4. The base of the logarithm is immaterial—the scale looks the same regardless of the base. Naturally the base will most probably be either 10 or e. We note that the distance between integral marks is not constant but lessens as we move from left to right. But the distance (measured with an ordinary ruler) is the same between 1 and 2 as between 2 and 4 (or between 4 and 8). To see why this is so, suppose we mark on the logarithmic scale x_1 and x_2 so that $x_2 = x_1/2$. Then $\log x_2 = \log x_1 - \log 2$, so that $\log x_1 - \log x_2 = \log 2$. Thus, the distance between the marked x_2 and x_1 on the logarithmic scale is constant, namely, $\log 2$. This occurs regardless of what x_2 and x_1 are as long as $x_2 = x_1/2$. In the same way, of course, 1, 3, 9, and 27 are the same distance apart.

We can construct a two-dimensional coordinate system in which one coordinate scale (either vertical or horizontal) is logarithmic and the other is *uniform*. Graph paper which uses such a system is called *semilog* paper. It could look like these:

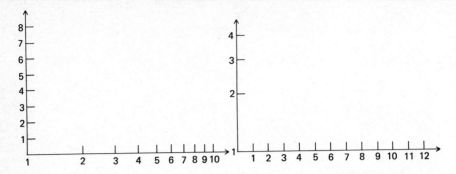

If *both* coordinate scales of a two-dimensional coordinate system are logarithmic, it is called *log* (or *log-log*). It could look like this:

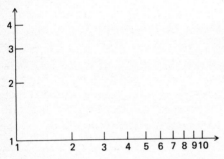

Note, once again, that a point whose coordinates are (5,7) on this type of graph paper is actually (log 5, log 7) numerically. What the logarithmic scales do, then, is to automatically take the logarithm of whatever number is plotted.

These coordinate systems are extremely useful on many occasions. We shall be presenting some applications to curvefitting, especially to curves of type $y = ab^x$ and $y = ax^n$.

First, consider the exponential curve with equation $y = ab^x$. We have already seen in Sec. 7.1 the graph of such a function using the ordinary uniform x and y coordinate system. Now suppose we take logarithms base b of both sides of the equation. Then, $\log_b y = \log_b(ab^x) = \log_b a + \log_b b^x = \log_b a + x$. That is, $\log_b y = \log_b a + x$. Now, let $y' = \log_b y$, $\alpha = \log_b a$, $x' = x$. Then we have the equation $y' = \alpha + x'$. This is the equation of a *straight line* with slope 1 in the (x',y') coordinate system. But, since $y' = \log_b y$, if the (x,y) coordinates of points lying on the curve $y = ab^x$ were to be plotted on semilog graph paper with the y axis logarithmic, the resulting curve would be a *straight line*.

Suppose $y = 5(2^x)$.   The table of values is

x	y
1	10
2	20
3	40
4	80
5	160

As before, the uniform scale graph appears thus:

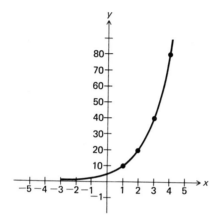

The graph on semilog paper looks like this:

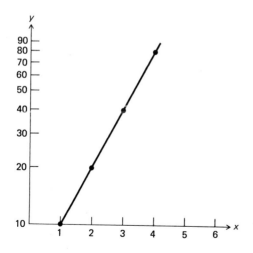

Now consider $y = ax^n$. Taking the logarithm of both sides of this equation (any base), we have $\log y = \log(ax^n) = \log a + \log x^n = \log a + n \log x$. Let $\log y = y'$, $\log a = \alpha$, $\log x = x'$. Then we have $y' = \alpha + nx'$. Again, we have arrived at the equation of a straight line. Now, however, $y' = \log y$ and $x' = \log x$, so that if the (x,y) coordinates of points lying on $y = ax^n$ were to be plotted on log-log paper (both scales logarithmic), the resulting curve would be a straight line.

Suppose $y = 5x^2$, $x > 0$. The ordinary graph is a portion of a parabola. On this curve lie the following points: $(1,5)$, $(2,20)$, $(3,45)$, $(4,80)$, $(5,125)$, $(6,180)$. We plot these points on log-log paper.

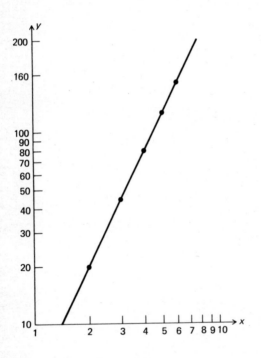

Semilog and log-log graph paper can often be of assistance in the process of fitting data to a curve. Recall that any set of data can be fitted to *any* curve, and the method of least squares will be of considerable use in finding the best-fitting curve of the type selected. We still have the important preliminary problem of selecting the type of curve the data should be fitted to. Obviously, data like the following would not be well fitted by a straight line:

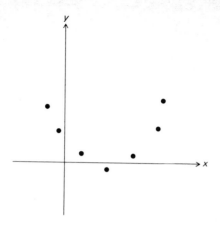

Even so, a straight line *can* be fitted, but the deviations of the line from the data points will be so large as to obviate any further practical use of the linear equation. For these data, one would do better to select a parabolic type of function like $y = ax^2 + bx + c$ since the data look, at least approximately, very like a parabola.

The practical way to proceed in this selection process is to first plot the data points on an ordinary uniform coordinate system. Glance casually at this plot and select from a list of suitably fitting curves one which seems to best fit the data. Then to get a very rough idea of the fitting equation, one could sketch in by hand a curve of that type and, using points on this rough sketch, find the equation of the curve. The equation would be very approximate, of course. The better idea would be, after selecting the type of curve to be fitted, to call upon the good process of the method of least squares to calculate the various parameters that occur in the equation. At least, one would be getting the "best" curve of that type.

We have already investigated fitting data to the straight line and to the parabola, so that if the plotted data (on uniform scale coordinates) appear either to be in a relatively straight line or to lie near a reasonable parabola, our choice and further procedure are clear. We can now consider fitting data to curves of two more types, $y = ae^{bx}$ and $y = ax^n$. Thus, if we were not pleased with the fitting to a line or parabola, we could proceed to plot the data on semi-log paper. Then if the data show up as nearly in a straight line, we could consider fitting the curve $y = ae^{bx}$. Again we could proceed very roughly from the straight line, drawn on the semilog plot, that approximates the data; perhaps from the slope of the line and from its intercepts we could find rough values for a and b. Or, more profitably, we could proceed directly to a modified version of the fitting of data to a straight line using the method of least squares to calculate a and b. We shall discuss this technique in a moment. Finally, we could plot the original data on log-log paper. If the points then appear to be approximately in a straight line, we would consider fitting a curve of the type $y = ax^n$.

Now before we present the method of linearizing mentioned above, we recall the following. Given the data

x'	y'
x'_1	y'_1
x'_2	y'_2
.	.
.	.

the method of least squares fits the data to the line with equation $y' = \alpha + \beta x'$ if α and β are determined by solving simultaneously the equations

$$\begin{cases} n\alpha + (\Sigma x')\beta = \Sigma y' \\ (\Sigma x')\alpha + (\Sigma x'^2)\beta = \Sigma x'y' \end{cases}$$

Suppose we are given the following data:

x	y
1	8.1
2	11.3
3	13.7
4	15.8
5	19.0

We discover that when these data are plotted on log-log paper, the points lie approximately in a straight line. We decide to fit the data to a curve of the type $y = ax^n$. Clearly, we must determine the parameters a and n. This is how we proceed to linearize the problem. We first take the log of both sides of $y = ax^n$. As before, we end with the linear equation $y' = \alpha + \beta x'$, where $y' = \log y$, $x' = \log x$, $\beta = n$, and $\alpha = \log a$. We decide to use natural logarithms. We want to use the method of fitting to a straight line that we have just reviewed. The data used in that method are values of x' and y' (not x and y). Thus, we must first construct a new set of data from the original set in x and y, recalling that $y' = \ln y$ and $x' = \ln x$. We construct the following new table:

$x' = \ln x$	$y' = \ln y$	$x'y'$	x'^2
0.000	2.092	0.000	0.000
0.693	2.425	1.681	0.482
1.099	2.617	2.876	1.208
1.386	2.760	3.825	1.921
1.609	2.944	4.737	2.589
Σ 4.787	12.838	13.119	6.200

We have used a table of natural logarithms and lengthy arithmetic. Substituting into the simultaneous equations for α and β, we have

$$\begin{cases} 5\alpha + 4.787\beta = 12.838 \\ 4.787\alpha + 6.200\beta = 13.119 \end{cases}$$

Solving these equations, we arrive at $\alpha = 2.08$, $\beta = 0.508$. Therefore, $n = \beta = 0.508$ and $a = e^{\alpha} = e^{2.08} = 8.00$. The equation of the fitting curve is $y = 8.00x^{0.508}$.

As we review this exercise, we realize that the process of linearizing can be used whenever the fitting curve can be rewritten in the form $y' = \alpha + \beta x'$. We then fit a straight line to the transformed data (x', y') and transcribe α and β back to the original parameters in the equation of the fitting curve. It is important to note that this is not the same process as applying the method of least squares directly to the fitting function.

We hope that you can figure out how to proceed if the data plotted on semilog paper turn out to be almost in a straight line. Other curves can be used for fitting purposes. If the method of linearizing is to work, such a function should contain only two parameters. Other examples are $y = ab^x$ and its special version $y = ae^{bx}$. Some types, like $y = a/(b + x)$ or $y = (x + a)/(y + b)$, are not suitable for the linearizing method. They can, however, be used following much more analysis with the general method of least squares.

EXERCISES

7.9 Use the method of this section to fit the following data to the more appropriate form of $y = ax^n$ and $y = ae^{bx}$:

(a)

x	y
1.02	4.5
2.72	14.9
7.39	49.5
20.10	163.0
55.00	521.0
148.00	1810.0

(b)

x	y
1.11	0.62
1.35	0.68
1.49	0.72
1.64	0.73
2.25	0.84
2.74	0.89

(c)

x	y
−1.0	0.05
−0.8	0.12
−0.4	0.57
0.2	0.63
0.7	47.00
1.2	330.00
1.3	517.00

7.10 Use the methods described in this section to fit the following data to the curve of type $y = ab^x$:

(a)

x	y
-3	5.64
-2	4.44
-1	3.48
0	2.83
2	1.78
3	1.42
4	1.14

(b)

x	y
0.1	6.69
0.4	9.39
0.6	10.60
0.8	12.90
1.0	18.00
1.1	20.10
1.3	24.60
1.5	32.40

7.5 APPROXIMATIONS BASED ON SERIES

As you have discovered, hand calculations involving exponential and logarithmic functions have required looking up the values of these functions in a table. The computer does not maintain such tables in its memory. Instead, it can calculate the value of these functions (and many others) almost instantaneously for any reasonable value of the argument with a great degree of accuracy. You simply write ALOG(X) or EXP(X), give the value of X and where you want the result stored, and the computer automatically calculates and stores that specific value of the function. One can, of course, write a program to have the computer calculate complete tables of functional values such as those that appear in handbooks. In this section we shall present some notions about the methods the computer uses to compute functional values such as these.

In the calculus course you may soon be taking, formulas for the calculation of the values of transcendental functions will be derived. Most of these formulas are based on the notion of finding a *power series* which represents the function and then truncating it appropriately to a finite polynomial. For our purposes the following discussion will have to suffice.

A *power series* in powers of $(x - a)$ is a "huge" polynomial of the form $b_0 + b_1(x - a) + b_2(x - a)^2 + b_3(x - a)^3 + \cdots + b_n(x - a)^n + \cdots$, or, if $a = 0$, in powers of x, $b_0 + b_1x + b_2x^2 + b_3x^3 + \cdots + b_nx^n + \cdots$. The b_i are (real) constants. As you note, the series extends indefinitely—it has an infinite number of terms. Now it turns out that for certain values of x, the series *converges;* that is, it has a sum. It is clear by inspection that every power series converges for $x = a$ and the sum is precisely b_0. In general, a power series converges in some interval $\alpha < x < \beta$. Outside that interval it does not converge, has no sum, and will be useless for our purposes.

Suppose that $\alpha < x < \beta$ is the *interval of convergence* of a given power series. Then for each x in the interval, the series has a unique sum. In a sense, the series is a function whose domain is the interval of convergence. Let us denote this function by $f(x)$. Thus, $f(x) = b_0 + b_1(x - a) + b_2(x - a)^2 + \cdots$, $\alpha < x < \beta$. We cannot in general find out what this function $f(x)$ is in explicit form, like $f(x) = e^x$ or $f(x) = 1/\sqrt{1 + x^2}$. Even so, we can still find the value of $f(x)$ for a value of x in the interval of convergence. Let us use an example to demonstrate the techniques involved.

Example. Let $f(x) = x + x^3/5 + x^5/25 + x^7/125 + \cdots + x^{2n+1}/5^n + \cdots$, $-\sqrt{5} < x < \sqrt{5}$. We calculate $f(0.2)$.

$$f(0.2) = 0.2 + \frac{(0.2)^3}{5} + \frac{(0.2)^5}{25} + \frac{(0.2)^7}{125} + \cdots$$

$$= 0.2 + \frac{0.008}{5} + \frac{0.00032}{25} + \frac{0.0000128}{125} + \cdots$$

$$= 0.2 + 0.0016 + 0.0000128 + 0.0000001024 + \cdots$$

$$\doteq 0.201613824$$

We note that as more terms are added on, the sum increases but at a much slower pace. We would suspect (and you can verify this) that adding on the next term of the series $(0.2)^9/625$ would not affect the first four figures of our present sum. In any case, we note that the number of terms that need be used to approximate $f(x)$ depends on *what* x is and *also* on the required degree of accuracy. It appears that in our example $f(0.2) \doteq 0.20$ to two significant figures [using the approximation $f(x) \doteq x$], $f(0.2) = 0.202$ to three significant figures [using the approximation $f(x) \doteq x + x^3/5$], $f(0.2) = 0.2016$ to four significant figures [using the approximation $f(x) \doteq x + x^3/5 + x^5/25$], and $f(0.2) = 0.20161$ to five significant figures (using the first four terms of the series). These approximations, like $f(x) \doteq x + x^3/5$, are simply truncations of the series and are simple polynomial approximations to $f(x)$.

Suppose we desire the value of $f(x)$ to be accurate to four decimal places. We could proceed like this. Start evaluating and adding the terms of the series one by one. After each addition, see if the absolute value of the *difference* between the new approximate sum and the immediately preceding one is less than 5×10^{-5}. When this occurs, the new sum would have to be accurate to four decimal places. In our example above, $f_1(0.2) = 0.2$, and $f_2(0.2) = 0.2 + 0.0016 = 0.2016$. Now $|f_2 - f_1| = 0.0016$, which is greater than 5×10^{-5}, so we proceed to the next term. $f_3(0.2) = 0.2016128$. Now $f_3 - f_2 = 0.0000128$, which is less than 5×10^{-5}. We stop and take 0.2016 as the result desired. Such a general procedure would obviously be amenable to computer programming.

7.11 Let $f(x)$ be defined by the following power series. Calculate $f(0.1)$ and $f(1.)$ correct to three decimal places:

$$f(x) = 1 + \frac{x}{2 \cdot 10} + \frac{x^2}{3 \cdot 10^2} + \frac{x^3}{4 \cdot 10^3} + \cdots + \frac{x^n}{(n+1) \cdot 10^n} + \cdots$$

7.12 Calculate $f(0.01)$ and $f(0.1)$ correct to five decimal places if

$$f(x) = x - \frac{x^3}{3} + \frac{x^5}{5} - \frac{x^7}{7} + \cdots + \frac{(-1)^n x^{2n+1}}{2n+1} + \cdots$$

7.13 Calculate $f(0.1)$ and $f(0.5)$ correct to four decimal places if

$$f(x) = 1 + x + \frac{x^2}{2!} + \frac{x^3}{3!} + \frac{x^4}{4!} + \cdots + \frac{x^n}{n!} + \cdots$$

(*Note:* Here we are using the *factorial*. We read $n!$ as "n factorial." By definition, $0! = 1$ and $n! = 1 \cdot 2 \cdot 3 \cdots n$ if n is an integer ≥ 1. Thus, $5! = 1 \cdot 2 \cdot 3 \cdot 4 \cdot 5 = 120$.)

7.14 Write a complete Fortran program to calculate

$$f(x) = 1 + x + \frac{x^2}{2!} + \cdots + \frac{x^n}{n!} + \cdots \qquad \text{for } 0 \leq x \leq 2$$

in intervals of 0.1, that is, for $x = 0, 0.1, 0.2, 0.3, \ldots, 1.9, 2.0$. Each value is to be correct to four decimal places. Print each x, the corresponding value of $f(x)$, *and* the value of e^x. Recall the Fortran library function EXP(X).

Now we consider the reverse situation which is also studied in the calculus. Suppose we are given an explicit function $f(x)$. What is the corresponding power series in powers of $(x - a)$, given a, and where does this series converge? We shall not be able to answer this question. We do give some important examples, however.

$$e^x = 1 + x + \frac{x^2}{2!} + \frac{x^4}{4!} + \cdots + \frac{x^n}{n!} + \cdots \qquad -\infty < x < \infty \qquad (7.4)$$

$$\ln x = (x - 1) - \frac{(x-1)^2}{2} + \frac{(x-1)^3}{3} + \cdots + \frac{(-1)^n (x-1)^n}{n} + \cdots$$

$$0 < x \leq 2 \qquad (7.5)$$

$$\sin x = x - \frac{x^3}{3!} + \frac{x^5}{5!} + \cdots + \frac{(-1)^n x^{2n+1}}{(2n+1)!} + \cdots \qquad -\infty < x < \infty \qquad (7.6)$$

$$\tan^{-1} x = x - \frac{x^3}{3} + \frac{x^5}{5} - \cdots + \frac{x^{2n+1}(-1)^n}{2n+1} + \cdots \qquad -1 \leqslant x \leqslant 1 \qquad (7.7)$$

After each series its interval of convergence is given. We have included series for the trigonometric functions $\sin x$ and $\tan^{-1} x$ to show how extensive the series representation of functions can be. You note that series (7.4) is precisely the one used in Exercise 7.14.

A polynomial can be obtained by truncating any of these power series after a suitable number of terms. This polynomial can then serve as an approximation to the function. As you recall, the number of terms one should use depends on the value of x to be substituted and how accurate the resulting approximation is required to be. It can be shown that at least 12 terms of series (7.4) will be required to achieve an accuracy of 10^{-8} if $-1 \leqslant x \leqslant 1$. This is not too distressing, but some series, although they surely converge, do it so slowly that a very much larger number of terms must be used to achieve accuracy like this. Recall that because of roundoff-error accumulation this could lead to an almost guaranteed severe loss of accuracy.

It would be very useful if there were more *condensed* series still representing $f(x)$ which would converge more rapidly, at least over a certain range of x, and still ensure the same accuracy. In fact, the computer makes use of such *economized* power series rather than the usual power series we have just given above. We cannot here present how they are derived, but we can point out some typical examples. Perhaps you will make a further study of this matter in a course in numerical analysis. For example, the series for $\ln(1 + x) =$ $x - x^2/2 + x^3/3 + x^4/4 + \cdots$, $-1 < x \leqslant 1$, which is obtained from series (7.5) by replacing x by $x + 1$, is very, very slow to converge. Many *hundreds* of terms would be required to achieve any reasonable accuracy, and even then the result would be suspect because of accumulated roundoff error. C. Hastings, Jr. in "Approximations for Digital Computers," Princeton University Press, 1955, has shown that the following polynomial approximates $\ln(1 + x)$ to an accuracy of 0.0000015 for *all* x in the interval $0 \leqslant x \leqslant 1$:

$$\ln(1 + x) \doteq 0.99990167x - 0.49787544x^2 + 0.31765005x^2$$
$$- 0.19376149x^4 + 0.08556927x^5 - 0.01833851x^6$$

Formulas like this are the sort that are actually used in the library function subprograms of Fortran to compute very efficiently functions like $\ln(1 + x)$ and all the other elementary functions like $\sin x$, $\cos x$, $\tan^{-1} x$, and e^x.

8

Boolean Algebra

8.1 INTRODUCTORY REMARKS

In this chapter we shall develop that application of Boolean algebra which is concerned specifically with the design and use of logic circuits for electronic computers. The general subject of Boolean algebra can be specialized in other ways so that it will apply to set theory and probability, mathematical logic, and switching circuits for telephone and control systems. This algebra, which in its original form was quite theoretical and formal, thus turns out to be quite practical and extremely useful.

In order to begin understanding the subject of Boolean algebra, we shall keep before us a very simple physical application—a simple on-off switch. The physical means used to activate or deactivate the switch will not concern us. It can be turned on or off manually or by means of an electromagnet. We shall let $\overset{A}{\circ\!\!-\!\!\|\!\!-\!\!\circ}$ stand for a simple switch (named A) which is either *on* so that current is able to flow through it—the variable A then has value 1—or *off* (or open so that current cannot flow through it)—the variable A then has value 0. A *switching circuit* may consist of several switches which we may name $A, B, C, \ldots$ arranged in series-parallel branches. We shall discuss these arrangements in the next section. In any case, at any specific time each of the switches will be either on or off. The name of each switch is a variable which may take on only one of the two values 0 or 1. We note, of course, that these values are precisely the two digits which are used in binary numbers.

Although we are using the simple switch as our physical realization of the Boolean variables $A, B, C, \ldots$, any device which operates in two distinct states can be used. Such devices are called *two-state devices*, and many of them are in current use. These include diodes, transistors, and other electronic components. In these, *on* and *off*, the two states, vary from one device to another. We thus have such states as on-off, closed-open, conducting-nonconducting, charged-discharged, magnetized-not magnetized, high voltage-low voltage, and so forth.

8.2 SERIES–PARALLEL SWITCHING CIRCUITS. LOGIC BLOCKS

Let us first consider a series circuit of two switches shown in Fig. 8.1. It seems obvious that current will flow from input to output only if both switches *A and B* are closed. Current cannot flow if either *A* or *B* is open or if they are both open.

Associated with each switching circuit is a *function* of the variables it contains. This function is called the *switching function*. The value of this Boolean function is to reflect the state of the *entire* circuit—its value will be 1 if the entire circuit is closed and 0 if the entire circuit is open. Clearly, the switching function depends upon the states of the various switches that compose it. Our first project is to figure out how to write the *switching function*.

For the series circuit shown in Fig. 8.1 we introduce the operation of multiplication, and we write the switching function as $f = A \cdot B$. We can read this function "A and B." From physical considerations we see that $f = 1$ only if both *A and B* are 1. It is 0 for all other combinations of states of the switches *A* and *B*. A table of functional values for f can be easily constructed. In this table we display the value of f for all possible combinations of states of the switches *A* and *B*. Recall that *A* and *B* can take on only values 0 or 1. We have the following table:

A	B	$f = AB$
0	0	0
0	1	0
1	0	0
1	1	1

Fig. 8.1

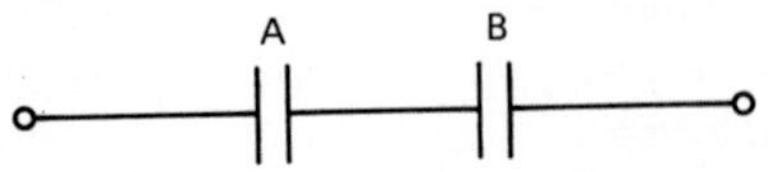

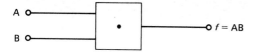

Fig. 8.2

We note that this multiplication table is exactly the same as for ordinary arithmetic multiplication of binary digits.

We now move naturally to the *logic block* corresponding to the same state of affairs. We introduce the AND block as shown in Fig. 8.2.

The AND block has two "wires" to carry in the two inputs and one "wire" to carry out one output. Signals appear at the two input terminals. They have "values" of either 0 or 1. In later applications we shall call these signals *binary* signals. Recall that we are idealizing an actual circuit, and actual signals could be, for example, a voltage of some measurable magnitude appearing at terminal A, none appearing at terminal B. We would say that $A = 1$, $B = 0$ at this time. The block contains the components (electronic or otherwise) that react to the two inputs and as a result output $A \cdot B$. Note that an output appears only if both A *and* B have signals. It is this property that leads us to call this block the *AND* logic block. The multiplication table given above for the series circuit switches clearly is to apply.

Consider now the parallel switching circuit shown in Fig. 8.3. Clearly, a current will be conducted through the circuit if either A *or* B is on or if they are both on. The only way current will not be conducted from input to output is if A and B are both off. For this circuit, we introduce the operation of addition, and we associate the switching function $f = A + B$. We can read this function "A or B." From physical consideration, the following addition table is evident:

A	B	f = A + B
0	0	0
0	1	1
1	0	1
1	1	1

Fig. 8.3

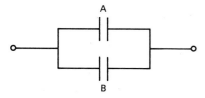

Addition appears to be normal except for the entry $1 + 1 = 1$. Although this is peculiar in ordinary arithmetic, it is evident that in Boolean arithmetic this statement simply means that in a parallel circuit if both switches are closed, then so is the circuit. We therefore accept this equation in Boolean arithmetic.

The logic block corresponding to the parallel switching circuit is the OR block which we denote:

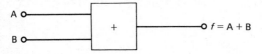

The OR logic block has two input "wires" and one output "wire." Again binary signals (0 or 1) appear at the input terminals, and the OR block circuit outputs the value of $f = A + B$ as given in the addition table above.

The third essential operation in Boolean algebra is that of *inversion*. We write A' and read "*not A*." The value of $f = A'$ is 0 if A is 1 and is 1 if A is 0. The table is simply

A	$f = A'$
1	0
0	1

This operation simply inverts the signal and is called *inversion*. The inversion logic block is

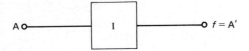

In a switching circuit we shall denote A' by

To review, in Boolean algebra there are two constants 0 and 1, any number of variables denoted by letters like $A, B, C, D, \ldots$, and three basic operations: multiplication (AND), addition (OR), and inversion (I). Recall that a variable can take on different values (either 0 or 1) at different times. At any given time it has one of the values 0 or 1. We note that 0 and 1 are the binary digits from which numbers in the binary system are constructed, and we are looking forward eagerly to see how logic blocks and other logic circuits can be used to store and manipulate binary numbers and possibly add them. Finally, physically a binary signal (digit) can be thought of as appearing on a "wire" according to the level of voltage which appears on the wire. We can let 1 stand for a higher (or more positive) voltage and 0 stand for a lower (or more negative) voltage. The logic circuit which constitutes the logic blocks (we shall not be able to design them) react to these input signals (voltages) and output a signal (voltage) according to their natures as described above.

8.3   BASIC OPERATIONS.   IDENTITIES.   THEOREMS

We have already accepted the following identities for the constants 0 and 1:

$$0' = 1$$
$$1' = 0$$
$$0 + 0 = 0$$
$$0 + 1 = 1 + 0$$
$$= 1 + 1$$
$$= 1$$
$$1 \cdot 1 = 1$$
$$0 \cdot 1 = 1 \cdot 0$$
$$= 0 \cdot 0$$
$$= 0$$

Since A and B can take on only values 0 or 1, it is clear that $AB = BA$ and $A + B = B + A$ for *any* variables A and B.   Thus, in Boolean algebra the operations of multiplication and addition are commutative.

Identities like those above are easily seen to be true.   In general, however, we have the very important property: A relation among variables is an *identity* if it holds for all possible combinations of the values of the variables it contains.

We make use of this property as we now extend our addition and multiplication to more than two variables.   We now show that $(A + B) + C = A + (B + C)$; that is, addition is associative and the parentheses directing the order of adding are unnecessary.   Consider the following "truth" table:

A	B	C	$A + B$	$(A + B) + C$	$B + C$	$A + (B + C)$
0	0	0	0	0	0	0
0	0	1	0	1	1	1
0	1	1	1	1	1	1
0	1	0	1	1	1	1
1	0	0	1	1	0	1
1	0	1	1	1	1	1
1	1	1	1	1	1	1
1	1	0	1	1	1	1

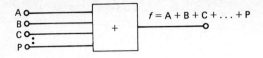

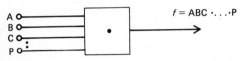

Fig. 8.4

The first three columns contain the eight possible combinations of values of the variables $A, B,$ and C. The column headed $A + B$ is found by simply adding the corresponding entries in the A and B columns using the basic addition table for the addition of two numbers. The column headed $(A + B) + C$ is found by adding the corresponding entries in the $A + B$ column and the C column. The other columns are found in exactly the same manner. Now we finally compare the entries in the columns headed $(A + B) + C$ and $A + (B + C)$ and we note that the columns are identical. Thus, $(A + B) + C = A + (B + C)$ is an identity. We have shown that addition is associative. In an entirely similar way we could show that $(AB)C = A(BC)$ so that multiplication is also associative. We could have reverted to the switching-circuit versions of these identities and convinced ourselves that they are.

In order to facilitate the addition and multiplication of more than two variables, we now extend the AND and OR logic blocks to allow any number of inputs greater than or equal to two, as shown in Fig. 8.4. Each of these blocks has one output. For the OR block $f = 1$ if any of $A, B, C, \ldots, P$ is 1, and $f = 0$ only if all $A, B, C, \ldots, P$ are 0. In the AND block $f = 0$ if any of $A, B, C, \ldots, P$ are 0, and $f = 1$ only if all $A, B, C, \ldots, P$ are 1.

We now present some examples of switching circuits and corresponding logic block diagrams which *realize* Boolean functions.

Example. $f = AB + C'$

THE SWITCHING CIRCUIT   THE LOGIC BLOCK DIAGRAM

We shall assume throughout that inverted values of variables are not available. This requires the use of the inverter to obtain C'.

Example. $f = A(AB + C + C'D)$

THE SWITCHING CIRCUIT

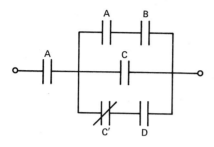

THE LOGIC BLOCK DIAGRAM

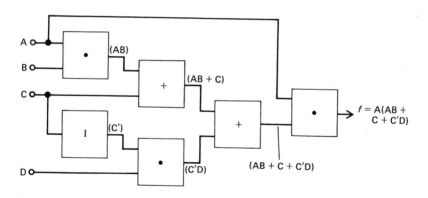

So that you may trace the development of the function, we have shown the output of each block. The presence of both C and C' in the switching circuit diagram means that a switch is in essence *ganged* or coupled from two switches that operate together but oppositely so that when one is open the other is closed. Note that we read the figures from left to right; thus arrows indicating the flow direction are not necessary.

Example. $f = ABCD' + C(AB + D)$

THE SWITCHING CIRCUIT

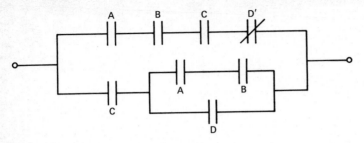

THE LOGIC BLOCK DIAGRAM

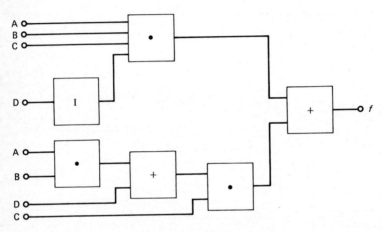

In the switching circuit the presence of two switches named A (or B) indicates that these two switches are ganged and act together simultaneously either both open or both closed. The block diagram shows two A, B, C, D inputs in order to simplify the geometry of the diagram. Actually, the signal on A appears simultaneously at both input wires labeled A. For these diagrams suppose that at a certain time $A = 1$, $B = 1$, $C = 1$, and $D = 0$. We find the value of f in different ways.

1. By use of the switching diagram, switches A, B, and C are closed and, since D is open, D' is also closed. This means that the entire upper branch is closed and clearly $f = 1$ (regardless of the condition of the lower branch).

2. In the logic block diagram, the output of the upper AND block is 1 since $A = B = C = D' = 1$. The lower first AND block has output 1, the OR block has output 1, and thus $f = 1 + 1 = 1$.

3. Using Boolean algebra and arithmetic, we have

$$f = ABCD' + C(AB + D)$$
$$= (1)(1)(1)(0') + 1(1 \cdot 1 + 0)$$
$$= 1 \cdot 1 \cdot 1 \cdot 1 + 1 \cdot 1 \cdot 1 + 1 \cdot 0$$
$$= 1 + 1 + 0$$
$$= 1$$

As noted above, identities in Boolean algebra can be verified by use of the truth table. Sometimes of course identities can be proved on the basis of previously proved or verified theorems. Before we present some of these proofs, let us call your attention to a very important fact concerning a certain duality involved in Boolean expressions. Let us first define the *dual* of a Boolean expression.

If, in any Boolean expression, + is replaced by ·, · is replaced by +, 0 is replaced by 1, and 1 is replaced by 0, the resulting expression is called the *dual* of the original expression. For example, the dual of $A + B' + 1$ is $A \cdot B' \cdot 0$; the dual of $A(B + C' + DE)$ is $A + [BC'(D + E)]$. Note that inversions like B' and C' are unaffected. Note also, in the second example, the retention of parentheses and the insertion of new parentheses when DE appears as $D + E$ in the dual. As a matter of notation we could write $du(f)$ to indicate the dual of the expression for f. Thus, if $f = AB' + CD$, then $du(f) = (A + B')(C + D)$.

Recall that an identity in Boolean algebra (as in most algebras) is an equation which is true for all values of the variables it contains. We have what is called the *principle of duality:* If $f = g$ is a Boolean identity, then $du(f) = du(g)$ is also an identity. We cannot prove this principle, so we shall accept it and notice its uses and consequences. This will mean that as soon as one identity has been verified or proved, its dual will automatically be assumed proved. The dual can be proved independently, too. As we shall soon see, $A + 0 = A$ is an identity. Thus, its dual $A \cdot 1 = A$ is also an identity.

As in trigonometry there are myriads of identities in Boolean algebra. We shall consider and prove only a few, selecting those that are basic or those that are more involved but which have some important applications. Let us first consider basic simple identities involving 0 and 1. We demonstrate the truth of the identity using switching circuits and then prove it using either truth tables or algebraic methods.

$$A + 0 = A$$
Dual: $A \cdot 1 = A$

THE SWITCHING DIAGRAM FOR A + 0

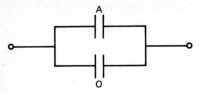

Here 0 represents a permanently *open* switch. Evidently, the circuit is open or closed as *A* is open or closed.

THE TRUTH TABLE METHOD

A	$f = A + 0$
0	0
1	1

Since the two columns are identical, $A + 0 = A$ is an identity. Recall that $0 + 0 = 0$ and $0 + 1 = 1$.

$$A \cdot 0 = 0$$
Dual: $A + 1 = 1$

THE SWITCHING CIRCUIT FOR A · 0

Here again 0 is a permanently open switch so that the circuit is always open whether or not *A* is closed. Thus, $A \cdot 0$ always has value 0.

THE TRUTH TABLE METHOD

A	$f = A \cdot 0$
0	0
1	0

$$A \cdot A = A$$
Dual: $A + A = A$

THE SWITCHING CIRCUIT FOR A · A

The two switches A act together so that if A is open, the circuit is open, and if A is closed, both switches are closed and the circuit is closed.

THE TRUTH TABLE METHOD

A	f = A · A
0	0
1	1

Recall that $0 \cdot 0 = 0$ and $1 \cdot 1 = 1$. We again note that this identity is not an algebraic equation. In Boolean algebra there is no such thing as $2A$ or A^2.

$$A + A' = 1$$
Dual: $A \cdot A' = 0$

THE SWITCHING CIRCUIT

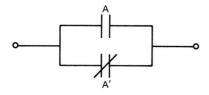

Regardless of the state of switch A, one branch or the other is always closed, and the circuit is thus always closed.

THE TRUTH TABLE METHOD

A	A'	f = A + A'
0	1	1
1	0	1

Recall that $0' = 1$ and $1' = 0$. Thus, $A + A'$ is always 1.

Now we turn to some important identities involving more than one variable. We note that as the number of variables increases, we rely more and more on algebraic and truth table methods. Switching circuits get too involved to use for the purpose of verifying identities.

$$A + AB = A \qquad \text{(Absorption law)}$$
Dual: $A(A + B) = A$

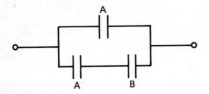

Notice that if A is closed, the circuit is closed regardless of the state of B. If A is open, the circuit is open, again regardless of the state of B.

THE TRUTH TABLE METHOD

A	B	AB	$f = A + AB$
0	0	0	0
0	1	0	0
1	0	0	1
1	1	1	1

Recall that to use a truth table with two variables, we must list every combination of possible values of the variables. There are four such combinations as shown. This law will be used quite often in the simplification procedures that we shall be eventually concerned with.

$$A(B + C) = AB + AC \quad \text{(distributive law)}$$
Dual: $A + BC = (A + B)(A + C)$

THE TRUTH TABLE METHOD

A	B	C	B + C	A(B + C)	AB	AC	AB + AC
0	0	0	0	0	0	0	0
0	0	1	1	0	0	0	0
0	1	0	1	0	0	0	0
0	1	1	1	0	0	0	0
1	0	0	0	0	0	0	0
1	0	1	1	1	0	1	1
1	1	0	1	1	1	0	1
1	1	1	1	1	1	1	1

Since the columns headed by $A(B + C)$ and $AB + AC$ are identical for every one of the eight possible combinations of values of A, B, and C, the distributive law is verified.   It is important to note carefully the dual of this law.   The "factorization" of $A + BC$ is different from ordinary algebra.   In simplification procedures we often use this dual identity.   Note that we can now prove the previous identity $A + AB = A$ algebraically.   Thus,

$$A + AB = A(1 + B) \qquad \text{(using the distributive law)}$$
$$= A \cdot 1$$
$$= A$$
$$A + A'B = A + B$$

Dual: $A(A' + B) = AB$

Proof: $A + A'B = (A + A')(A + B) \qquad \text{(factoring)}$
$$= 1 \cdot (A + B)$$
$$= A + B$$

Example. Simplify algebraically the expression $A + ABC + A'B'$.

$$A + ABC + A'B' = A(1 + BC) + A'B'$$
$$= A + A'B'$$
$$= A + B'$$

Example. Simplify algebraically the expression $B' + AB + BC$.

$$B' + AB + BC = (B' + A)(B' + B) + BC$$
$$= B' + A + BC$$
$$= B' + BC + A$$
$$= (B' + B)(B' + C) + A$$
$$= B' + C + A$$

The two preceding examples are given to show how some of the identities already covered can be used to simplify Boolean expressions.   In Sec. 8.5 we present a much more satisfactory method for the simplification process, one which does not require complicated selective choices of identities but which instead automatically uses identities by means of a semigraphical procedure.   This is the method of Karnaugh maps.

Now we present some more identities concerning inversion.   First, we want to allow expressions to be inverted such as $(A + B' + C)'$ and we must see how to do this. We already know that $0' = 1$ and $1' = 0$. We also have

$(A')' = A$ (double inversion)   (no dual)

THE TRUTH TABLE METHOD

A	A'	A''
0	1	0
1	0	1

$(AB)' = A' + B'$ (De Morgan's law)
Dual: $(A + B)' = A'B'$

THE TRUTH TABLE METHOD

A	B	AB	(AB)'	A'	B'	A' + B'
0	0	0	1	1	1	1
0	1	0	1	1	0	1
1	0	0	1	0	1	1
1	1	1	0	0	0	0

Since the entries in the column headed $(AB)'$ and those in the column headed $A' + B'$ are identical for every set of values of A and B, De Morgan's law, $(AB)' = A' + B'$, is an identity. It and its dual are very important and useful identities. The laws naturally extend to finding the inverse of the sum of any number of variables or the inverse of the product of any number of variables. For example, $(A + B + C + D + E)' = A'B'C'D'E'$ and $(ABCD)' = A' + B' + C' + D'$. Using De Morgan's law and the distributive law, one can multiply out a complicated expression like $[C(A + D)' + E']'$ into a simple sum form. Thus,

$$[C(A + D)' + E']' = (CA'D' + E')'$$
$$= (CA'D')'(E')'$$
$$= (CA'D')'E$$
$$= (C' + A + D)E$$
$$= AE + DE + C'E$$

Let us momentarily return to the subject of simplification of Boolean expressions by means of a physical example. Suppose we are given the following switching circuit:

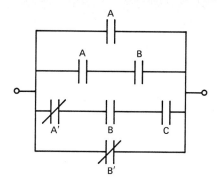

Clearly, the switching function is $f = A + AB + A'BC + B'$, and seven switches are needed. We use Boolean identities to simplify the switching function and thus, we hope, arrive at a completely equivalent function and corresponding switching circuit. Now

$$f = A + AB + A'BC + B'$$
$$= A + A'BC + B'$$
$$= A + BC + B'$$
$$= A + B' + BC$$
$$= A + B' + C$$

The equivalent switching circuit is

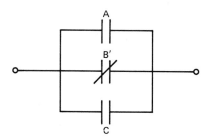

Now only three switches are needed, a saving of four switches. This is a reduction in cost of whatever physical element the switches represent. Recall that they may be mechanical switches, diodes, transistors, or other types of devices, and still we represent their operation, circuit, and simplification as we have done for simple, manually operated on-off switches.

The process of simplifying a function to its absolute minimum form is a major and most important phase in design. However, often several equivalent minimum forms result. If the expression to be simplified is not too involved and rather obvious simplifications come to mind without too much manipulation and deep thought, they should obviously be done algebraically. If this is not the case, some systematic way of simplifying Boolean expressions must be used. Of several methods in current use, we have selected the Karnaugh map method. But before we can present this method in detail, some preliminary groundwork must be laid. This is done in the following section.

EXERCISES

8.1 Draw switching circuits which realize each of the following Boolean functions:

(a) $f = A + B'C$
(b) $f = (A + B')(C + D) + A'$
(c) $f = (ABCD + E + F)G$
(d) $f = (AB' + A'B + AB)(A + B + C)$
(e) $f = D(A' + B' + C') + E(C' + D') + F$

8.2 Draw logic block diagrams using $\boxed{+}$, $\boxed{\cdot}$, and $\boxed{I}$ which realize each of the Boolean functions of Exercise 8.1.

8.3 Draw logic block diagrams which realize each of the following Boolean functions. Do not simplify first.

(a) $[(A + B)' + C]' + D$
(b) $(A'B' + C + D)(A + B)'$
(c) $(A + B + C + D + E)(A' + B' + C' + D')$
(d) $(AB + A'B')'(A + B')$
(e) $[(A + B')' + (C' + D)' + E]'$

8.4 Write the switching function f for each of the following switching circuits:

(a)

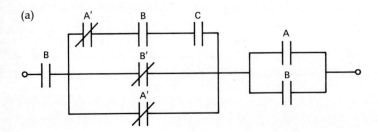

(b)

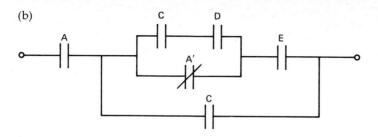

(c)

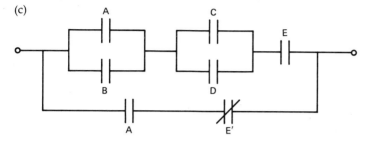

8.5 Draw the logic block diagram corresponding to each of the switching circuits of Exercise 8.4.

8.6 Simplify the following using *algebraic* techniques:

(a) $A(A' + AB)$
(b) $ABC + AB'C + A'B'C$
(c) $C(A + B'C')'$
(d) $(A + C)(A' + C')$
(e) $(AB' + A'B)'$
(f) $A + A'B + B'C$
(g) $A + B + C + A'B$
(h) $AB' + C(A' + B' + C)$
(i) $(AB)' + (A'BC)' + A$
(j) $(C' + D' + A')(A + C + D)$
(k) $B + AC + AB + C$
(l) $ABC' + E + D(ABC' + E)$

8.7 Use the truth table method to verify the following identities:

(a) $(A + B)(A' + B) = B$
(b) $B(B + C) = B$
(c) $A(A + B + C) = A$
(d) $(A + B' + C')' = A'BC$

(e) $(A' + AB)' = AB'$

(f) $(C + D)' + C = C + D'$

(g) $(A + B + C)(A' + B' + C) = AB' + A'B + C$

8.4 NORMAL FORMS FOR BOOLEAN EXPRESSIONS

An expression containing n variables is said to be in *sum normal form* if it is the sum of terms each of which contains each of the n variables (in inverted form or not).

Example. The expression $AB + A'B'$ is in sum normal form in the two variables A and B. But, clearly, the expression $A + AB$ is not. Also, $ABC + AB'C + A'BC'$ is in sum normal form in the three variables A, B, and C. $(A + C)' + BC$ is not. First we show how any expression can be put into sum normal form. The importance and the uses of the form will become evident as we proceed. Several examples follow which reveal the procedure.

Example. Place the expression $A + B$ in sum normal form in the variables A and B.

We note that the first term A does not contain the variable B. But $A = A \cdot 1 = A(B + B') = AB + AB'$ is equivalent to A and contains terms each of which has both of the variables A and B. Similarly, $B = (A + A')B = AB + A'B$. The result is $A + B = AB + AB' + AB + A'B = AB + AB' + A'B$, the sum normal form.

Example. Place the expression $A + B$ in sum normal form in the *three* variables A, B, and C.

$A = A \cdot 1 \cdot 1$
$\quad = A(B + B')(C + C')$
$\quad = ABC + ABC' + AB'C + AB'C'$

$B = B \cdot 1 \cdot 1$
$\quad = B(A + A')(C + C')$
$\quad = ABC + ABC' + A'BC + A'BC'$

Thus,

$A + B = ABC + ABC' + AB'C + AB'C' + A'BC + A'BC'$

Example. Place the expression $(A' + B)' + BC'$ in sum normal form in the three variables A, B, and C.

Applying De Morgan's law, we have $AB' + BC'$. Then

$$AB'(C + C') + (A + A')BC' = AB'C + AB'C' + ABC' + A'BC'$$

Consider the expression $AB + AB' + A'B + A'B'$. The sum contains *all possible* two-variable terms in A and B. It is called the *complete* sum normal form in two variables A and B. We note that $AB + AB' + A'B + A'B' = A(B + B') + A'(B + B') = A + A' = 1$. In general, the complete sum normal form in n variables has 2^n terms and *always* equals 1. Thus, the complete sum normal form in three variables A, B, and C has eight terms and their sum $ABC + ABC' + AB'C + AB'C' + A'BC + A'BC' + A'B'C + A'B'C'$ is 1.

Let us look at the terms of this last complete sum normal form. Suppose we select a particular combination of values for the variables, say $A = 0$, $B = 1$, $C = 0$. Evaluating these eight terms in order, we then have $ABC = 0(1)(0) = 0$, $ABC' = 0(1)(1) = 0$, $AB'C = 0(0)(0) = 0$, $AB'C' = 0(0)(1) = 0$, $A'BC = 1(1)(0) = 0$, $A'BC' = (1)(1)(1) = 1$, $A'B'C = 1(0)(0) = 0$, and $A'B'C' = 1(0)(1) = 0$. Exactly one term has value 1, the term $A'BC'$. All the rest have value 0. One can easily find the term which will be 1 for any other combination of values of A, B, and C by noting that if a variable is 0, we select the term containing that variable inverted, and if a variable is 1, we select that term which has that variable not inverted. Thus, should $A = 1$, $B = 0$, $C = 1$, then the term $AB'C$ is the one and only term in the complete sum normal form which will have value 1. This works for any number of variables. In four variables A, B, C, D the complete sum normal form has 16 terms, and the term $A'BCD'$ is the only one which equals 1 when $A = 0$, $B = 1$, $C = 1$, and $D = 0$.

It begins to be apparent that a function of, say, three variables A, B, and C is completely determined when one knows for which combinations of sets of values for A, B, and C the function has value 1. For example, suppose that f is a function of A, B, and C and has value 1 for each set in the following table and has value zero *for all other combinations*

A	B	C
1	0	0
0	1	0
0	1	1

Then $f = AB'C' + A'BC' + A'BC$.

Let us check up on this assertion. Clearly for each combination of values given in the table, f has value 1 since for each combination one of the terms of f has value 1 and the other two have value 0. Thus, the sum always equals 1. Also, for all the other eight combinations of values of A, B, and C each of the terms of f has value 0 and thus f has value 0.

Alternately, if a function is written in sum normal form, its truth table is evident. For example, suppose $f = A'BC + ABC' + A'B'C'$. Then

A	B	C	f
0	1	1	1
1	1	0	1
0	0	0	1

and f is 0 otherwise.

An interesting note: If f is written in sum normal form, then f' consists of those terms of the *complete* sum normal form which are not present in f. Thus, if $f = AB' + A'B + A'B'$, then $f' = AB$. You may check this by other means. If the values of f are read from a truth table, then f' consists of those terms for which f has value 0.

There is the dual normal *product* form in which the function is represented as the *product* of factors, each of which contains the sum of each of the variables in either inverted or noninverted form. For example, $(A + B)(A' + B')$ is in product normal form in the two variables A and B; $A + B'(A + B)$ is not. Similarly, $(A + B + C')(A' + B' + C')(A + B' + C)$ is in product normal form; $A + BC'$ is not. Since we shall not be making use of this form, which is equivalent to the sum form for a given function f, we shall not discuss its properties further.

EXERCISES

8.8 Place each of the following functions in sum normal form in two variables, A and B:

(a) A
(b) $(A' + B)'$
(c) $A + B'$

8.9 Place each of the functions in Exercise 8.8 in sum normal form in *three* variables, A, B, and C.

8.10 Place each of the following functions in sum normal form in three variables, A, B, and C:

(a) $B + C'$
(b) $A + B'C$

(c) $A'B + B'C$

(d) C

8.11 Place each of the following functions in sum normal form in four variables, A, B, C, and D:

(a) $BC' + D$

(b) $ABC + C'D$

(c) $A + BCD'$

(d) $(AB'C)'$

(e) $A + B + C + D'$

8.5 SIMPLIFICATION OF BOOLEAN EXPRESSIONS: KARNAUGH MAPS

One of the most used methods to simplify Boolean functions is the *Karnaugh map*. This method provides a definite procedure, a specific way, to simplify any Boolean expression.

Recall that if f is a function of two variables, we can display it as in the following truth table:

A	B	f
0	0	1
0	1	0
1	0	0
1	1	1

that is, $f = A'B' + AB$.

The Karnaugh map for this function is as follows:

A\B	0	1
0	1	
1		1

Note the $2^n = 4$ squares to allow for the four possible terms in the complete sum normal form for a function of two variables. The values of A are read across the top, B down the left. The upper block is thus the space for $A = 0$, $B = 0$. We place a 1 in those blocks corresponding to those pairs of values of A and B for which f has value 1. It is common practice to assume that for all blocks left empty the function has value 0.

For three variables we shall require a map with $2^3 = 8$ blocks. The standard form is as follows:

```
    A
BC \  0   1
0 0 ┌───┬───┐
    │   │   │
0 1 ├───┼───┤
    │   │   │
1 1 ├───┼───┤
    │   │   │
1 0 ├───┼───┤
    │   │   │
    └───┴───┘
```

Again it is clear that the importance of the sum normal form emerges. In order to fill in the appropriate squares to represent a function, the function must first have been expressed in sum normal form. Each block of the map represents a certain term in the complete sum normal form. A 1 is placed in each block which corresponds to the occurrence of a term of the sum normal form of f.

For four variables, 16 blocks are necessary. The map is as follows:

```
     AB
CD \   0 0   0 1   1 1   1 0
0  0 ┌─────┬─────┬─────┬─────┐
     │     │     │     │     │
0  1 ├─────┼─────┼─────┼─────┤
     │     │     │     │     │
1  1 ├─────┼─────┼─────┼─────┤
     │     │     │     │     │
1  0 ├─────┼─────┼─────┼─────┤
     │     │     │     │     │
     └─────┴─────┴─────┴─────┘
```

Although we shall not use them here, Karnaugh maps can be constructed for functions of five or six variables rather easily. Here is the map for five variables:

```
  A              0                          1

   \  BC                       \  BC
   DE \  0 0  0 1  1 1  1 0     DE \  0 0  0 1  1 1  1 0
   0  0                         0  0
   0  1                         0  1
   1  1                         1  1
   1  0                         1  0
```

One can very easily read out the unsimplified sum normal form for f from any of these maps.

Example

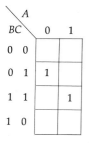

$f = A'B'C + ABC$

Example

AB CD		0 0	0 1	1 1	1 0
0	0		1		
0	1				1
1	1	1			
1	0				

$f = A'BC'D' + AB'C'D + A'B'CD$

We have two basic problems: (1) Reading into the map any function no matter what form it is originally in, and (2) reading out the *most simplified* equivalent form for f. Let us consider the first problem, first for functions of three variables. Any function of three variables after having been freed of all parentheses by the use of De Morgan's law, the double complementation law, and the distributive law, will then contain terms of three kinds: Terms which contain only one variable, terms which contain two variables, and terms which contain all three variables. For each of the terms containing three variables we simply mark a 1 in the corresponding square of the map. For the two- and one-variable terms we could proceed algebraically, expanding them to three-variable terms, that is, placing f in sum normal form. Thus, if the term is AB, we could write $AB = AB(C + C') = ABC + ABC'$ and then place a 1 in each of the squares corresponding to ABC and ABC'. *However, we want* the Karnaugh map procedure to be semiautomatic, and so instead of performing the algebra, we follow this rule: Place a 1 in each block for which $A = 1$, $B = 1$, regardless of what C is.   The map would then look like this:

BC \ A	0	1
0 0		
0 1		
1 1		1
1 0		1

Similarly, for a term like B, we place a 1 in each block where B is 1, regardless of what A or C is in those blocks.   The map for the term B would be

BC \ A	0	1
0 0		
0 1		
1 1	1	1
1 0	1	1

Example. Map the function $f = A + A'B'C'$.

For the term A we place a 1 in every block where A is 1, that is, in every block of the second column.   For the term $A'B'C'$, we place a 1 in the block in which $A = 0$, $B = 0$, and $C = 0$.   The map is

BC \ A	0	1
0 0	1	1
0 1		1
1 1		1
1 0		1

We note that a three-variable term requires only one block to display it, a two-variable term requires two blocks, and a one-variable term requires four blocks.

Now let us turn to the mapping of functions of four variables.   As before, the function will have terms involving one, two, three, or four variables.

The four-variable terms will require just one square, the three-variable terms will be displayed in two squares, the two-variable terms will require four squares, and the one-variable terms will require eight squares for display. For example, the term A is read in by placing a 1 in every square in which A is 1, regardless of what B, C, or D is. The map will appear thus:

CD \ AB	0 0	0 1	1 1	1 0
0 0			1	1
0 1			1	1
1 1			1	1
1 0			1	1

The term ACD will be read in by placing a 1 in two blocks, the two blocks for which $A = 1$, $C = 1$, $D = 1$, regardless of what B is. Now consider the function $f = A + CD$. It maps as follows:

CD \ AB	0 0	0 1	1 1	1 0
0 0			1	1
0 1			1	1
1 1	1	1	1	1
1 0			1	1

We note that the two terms generate 1s in some of the same squares. A *single* 1 is placed in the squares where they overlap. In the example above, the A term causes the eight 1s appearing in the third and fourth columns; the CD term causes the four 1s appearing in the third row. They overlap in the third and fourth columns.

Example. Plot the function $f = AB + CD' + A'B'C'D + AC'D$.

CD \ AB	0 0	0 1	1 1	1 0
0 0			1	
0 1	1		1	1
1 1			1	
1 0	1	1	1	1

You should be able to verify the entries in this table; note once again the overlapping of entries. The following exercise set gives you the opportunity to sharpen your skill in plotting Boolean functions on Karnaugh maps.

8.12 On a two-variable Karnaugh map, plot each of the following functions:

(a) $f = A$
(b) $f = A + B'$
(c) $f = AB' + A'B$
(d) $f = AB + A' + B'$

8.13 Plot each of the following functions on a three-variable map:

(a) $f = A + B$
(b) $f = A + BC'$
(c) $f = ABC' + A'C'$
(d) $f = A + B' + C$
(e) $f = AB + A'C + BC'$

8.14 Plot each of the following functions on a four-variable map:

(a) $f = A + D$
(b) $f = AB + C'D + AB'C$
(c) $f = ABCD' + A' + B'C$
(d) $f = A + B + C + B'D + CD'$
(e) $f = ABCD' + AB'C'D + A'BC'D$

8.15 Plot each of the following functions on a five-variable map:

(a) $f = C + D + E'$
(b) $f = AE + CDE$
(c) $f = ABCE' + A'D + C'$
(d) $f = E + CD' + ABC + AB'CDE'$
(e) $f = ABC'DE + AB'C'D'E' + A'B'CD'E$

We now turn to the essential part of this section—the reading out of the most *simplified* form of a function that has been plotted on a Karnaugh map. This is equivalent, of course, to simplifying a given function, a process most important in logic design.

Given a map, we proceed to "loop" it so that all the 1s it contains are enclosed in 1-loops, 2-loops, 4-loops, 8-loops, etc.

A 1-loop is a loop containing just one 1. For example,

We read out $f = AB$.

A 2-loop is a loop containing two 1s which are adjacent in one of the following senses: They are next to each other horizontally; they are next to each other vertically; or they lie at the ends of the same row or column. For example,

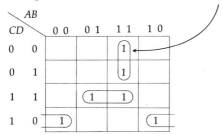

Notice that when two 1s are adjacent, the terms corresponding to the 1s differ in just one variable. For the marked two-loop, the terms are $ABC'D'$ and $ABC'D$—they differ in the D's only. Note that $ABC'D' + ABC'D = ABC'(D + D') = ABC'$ so that this looped pair of 1s is read out as ABC' (which contains the same variables which *did not change* in the two adjacent squares). For the bottom loop, the two terms are $A'B'CD'$ and $AB'CD'$. Thus, they add up to $B'CD'$. We note that the read-out of a two-loop always contains one less variable than the total number of variables of the map.

A 4-loop contains four 1s which are *adjacent* in the sense that the terms corresponding to the four 1s contain *two* variables which remain constant. Here are some examples of four-loops.

AB CD	0 0	0 1	1 1	1 0
0 0			1	
0 1			1	
1 1			1	
1 0			1	

Here A and B remain constant. The read-out is $f = AB$.

CD \ AB	0 0	0 1	1 1	1 0
0 0				
0 1				
1 1				
1 0	(1	1	1	1)

Clearly, C and D' remain constant. Hence, $f = CD'$. Let us check this read-out algebraically. First,

$$f = A'B'CD' + A'BCD' + ABCD' + AB'CD'$$
$$= (A'B' + A'B + AB + AB')CD'$$
$$= (1)CD'$$
$$= CD'$$

CD \ AB	0 0	0 1	1 1	1 0
0 0				
0 1				
1 1	1			1
1 0	1			1

Here B' and C remain constant. $f = B'C$.

CD \ AB	0 0	0 1	1 1	1 0
0 0			1	1
0 1			1	1
1 1				
1 0				

$f = AC'$

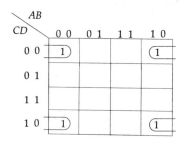

$f = B'D'$

Note that the example below is not a 4-loop; however, it is two 2-loops. Thus, $f = B'C'D' + B'CD$.

$f = B'C'D' + B'CD$

It is clear that the read-out of a 4-loop contains two less variables than the total number of variables in the map.

An 8-loop (in a map containing three or more variables) is a set of eight 1s which are adjacent in the sense that the terms corresponding to the 1s have *one* variable which remains constant. Here are some examples.

$f = D$

CD \ AB	0 0	0 1	1 1	1 0
0 0	1			1
0 1	1			1
1 1	1			1
1 0	1			1

$f = B'$

The read-out of an 8-loop contains three less variables than the total number of variables in the map.

Note that the following is *not* an 8-loop:

CD \ AB	0 0	0 1	1 1	1 0
0 0	1	1	1	1
0 1				
1 1	1	1	1	1
1 0				

It does contain two 4-loops so that the read-out is $f = C'D' + CD$.

We have so far presented functions whose maps contain single two-, four-, or eight-loops. Several of the maps contained several 2-loops or 4-loops. We now turn to maps that in general contain several loops of the same or of different sizes. We must learn how to most efficiently read out f. We note immediately that the larger the loop one can use, the better. We shall restrict ourselves to maps with at most four variables.

The general rules of procedure go like this:

1. Loop all 8-loops.
2. Proceed to the remaining uncovered 1s. Loop them with 4-loops as long as possible. An important note: It is perfectly permissible to have loops overlap; in fact, in many circumstances this is precisely the procedure to follow in order to ensure that the most simplified version of f will be read out.
3. Cover as many of the remaining 1s as possible with 2-loops.
4. Cover the rest of the 1s with 1-loops.

Then, following the read-out schemes given above, proceed to read out the terms represented by the loops. f is the sum of such terms, and if the loops have been judiciously applied as prescribed in these rules, the resulting f is the most simplified equivalent version of the function. Some examples illustrating the use of these general rules follow.

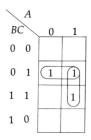

No 8-loops, no 4-loops, two 2-loops.
$f = B'C + AC$

Note the overlapping of the two 2-loops. It would not be correct to encircle the two right 1s in a 2-loop and the left 1 in a single 1-loop. If one did this, $f = AC + A'B'C$ could be simplified further: $AC + A'B'C = C(A + A'B') = C(A + A')(A + B') = C(A + B') = B'C + AC$, as before.

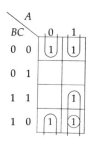

No 8-loops, one 4-loop, one 2-loop.
$f = C' + AB$

Again note the overlapping of the 2-loop and the 4-loop.

No 8-loops, no 4-loops, one 2-loop, one 1-loop.
$f = BC + A'B'C'$

CD \ AB	0 0	0 1	1 1	1 0
0 0	1	1		
0 1	1	1		
1 1		1	1	
1 0				

No 8-loops, one 4-loop, one 2-loop.
$f = A'C' + BCD$

CD \ AB	0 0	0 1	1 1	1 0
0 0	1	1	1	1
0 1		1		
1 1				
1 0	1			1

No 8-loops, two 4-loops, one 2-loop.
$f = B'D' + C'D' + A'BC'$

Again note the neat overlapping of loops.

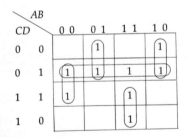

CD \ AB	0 0	0 1	1 1	1 0
0 0		1		1
0 1	1	1	1	1
1 1	1		1	
1 0			1	

No 8-loops, one 4-loop, four 2-loops.
$f = A'B'D + A'BC' + AB'C' + ABC + C'D$

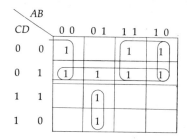

CD \ AB	0 0	0 1	1 1	1 0
0 0	1		1	1
0 1	1	1	1	1
1 1		1		
1 0		1		

No 8-loops, three 4-loops, one 2-loop.
$f = A'BC + AC' + B'C' + C'D$

CD \ AB	0 0	0 1	1 1	1 0
0 0		1	1	1
0 1		1	1	
1 1	1	1	1	
1 0	1	1	1	

One 8-loop, one 4-loop, one 2-loop.
$f = B + A'C + AC'D'$

We now present several applications of the reading in and reading out of functions. Suppose we read in the unsimplified function $f = A + BC'$ $+ A'BC' + A'B'$ and then read out the simplified version of f. To show clearly once again how the function is read in, we present the following Karnaugh maps:

BC \ A	0	1
0 0		1
0 1		1
1 1		1
1 0		1

+

BC \ A	0	1
0 0		
0 1		
1 1		
1 0	1	1

+

BC \ A	0	1
0 0		
0 1		
1 1		
1 0	1	

+

BC \ A	0	1
0 0	1	
0 1	1	
1 1		
1 0		

=

BC \ A	0	1
0 0	1	1
0 1	1	1
1 1		1
1 0	1	1

Normally, of course, the separate terms of f are read in directly on the final map of this sequence—here we are displaying each term's map for initial clarity. Then, the read-out of f from the last map is $f = A + B' + C'$.

As the final example, suppose function $f = AD' + A'CD + A'BC'D + D'$ is to be simplified and then a logic circuit is to be drawn which will realize the simplified function. First, we read in f as shown:

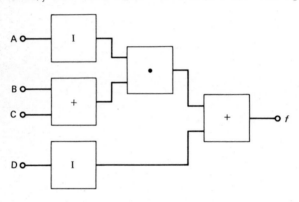

AB CD	0 0	0 1	1 1	1 0
0 0	1	1	1	1
0 1		1		
1 1	1	1		
1 0	1	1	1	1

Thus, $f = D' + A'B + A'C = D' + A'(B + C)$. The logic diagram appears thus:

EXERCISES

8.16 Use the looping method to read out the (most) simplified version of f if its Karnaugh map is:

(a)

BC \ A	0	1
0 0		1
0 1	1	
1 1	1	
1 0	1	1

(b)

BC \ A	0	1
0 0	1	1
0 1	1	1
1 1	1	
1 0		

(c)

BC \ A	0	1
0 0	1	
0 1		1
1 1	1	1
1 0		

8.17 Use the looping method to read out the simplified version of f if its Karnaugh map is:

(a)

CD \ AB	0 0	0 1	1 1	1 0
0 0	1			1
0 1		1	1	
1 1		1	1	
1 0	1			1

(b)

CD \ AB	0 0	0 1	1 1	1 0
0 0	1	1	1	1
0 1		1	1	
1 1		1		
1 0		1		

(c)

CD \ AB	0 0	0 1	1 1	1 0
0 0	1			1
0 1	1	1	1	1
1 1	1	1	1	1
1 0	1			

(d)

CD \ AB	0 0	0 1	1 1	1 0
0 0	1			
0 1	1		1	1
1 1	1			
1 0				1

(e)

CD \ AB	0 0	0 1	1 1	1 0
0 0		1	1	1
0 1		1	1	
1 1	1	1		
1 0	1			1

(f)

CD \ AB	0 0	0 1	1 1	1 0
0 0	1		1	
0 1	1		1	
1 1	1	1	1	1
1 0			1	

(g)

CD \ AB	0 0	0 1	1 1	1 0
0 0		1	1	1
0 1			1	1
1 1				1
1 0	1			

(h)

CD \ AB	0 0	0 1	1 1	1 0
0 0	1		1	
0 1	1	1	1	
1 1			1	
1 0			1	

(i)

CD \ AB	0 0	0 1	1 1	1 0
0 0	1	1		1
0 1		1		
1 1			1	1
1 0				1

(j)

CD \ AB	0 0	0 1	1 1	1 0
0 0	1			
0 1	1	1	1	
1 1	1	1	1	
1 1	1			1

8.18 Recall that f' can be read out from the Karnaugh map for f by reading out those blocks that contain 0s. Follow the method of looping but loop the 0s to read out f' for each of the Karnaugh maps of Exercise 8.17.

8.19 Read each of the following Boolean functions into a three-variable Karnaugh map (without algebraically simplifying). Then read out simplified f.

   (a) $f = A + BC' + A'BC$
   (b) $f = ABC + AB'C' + A'B + A$
   (c) $f = BC' + AB + A'BC' + AB'C$
   (d) $f = B + C + A'BC'$

8.20 Read each of the following Boolean functions into a four-variable Karnaugh map and then read out simplified f.
   (a) $f = ABC' + CD + AC'D + D'$
   (b) $f = A + B' + BC' + A'BC'D$
   (c) $f = A'BC + ABC' + A'C' + B'C'$
   (d) $f = A + BC + A'B'C + ABCD$

8.21 Write the switching function directly for each of the following switching circuits. Read f into an appropriate Karnaugh map, read out simplified f, and draw the corresponding simplified switching circuit.

(b)

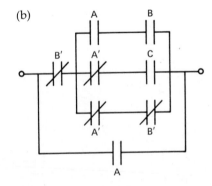

(a)

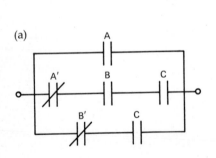

(c)

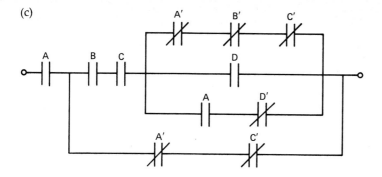

8.6   OTHER LOGIC BLOCKS: NAND, NOR

We have seen that any Boolean function can be realized (that is, constructed) from the three elemental logic blocks AND, OR, and INVERTER. In fact, however, just the two blocks AND and INVERTER are sufficient. To show that this is so, we make good use of De Morgan's law: $(A'B')' = (A')' + (B')' = A + B$. Thus,

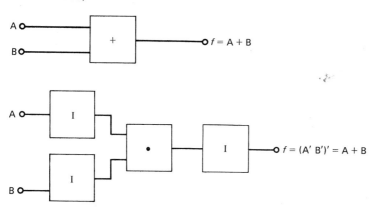

and $A + B$ can be constructed using the operations of $\cdot$ and inversion. In a similar way, the blocks OR and INVERTER are sufficient, since $(A' + B')' = (A')'(B')' = AB$, so that

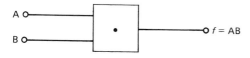

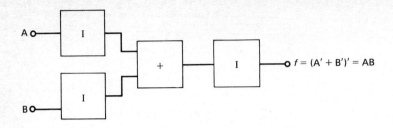

$f = (A' + B')' = AB$

A set of blocks which can implement any Boolean function is said to be *functionally complete*. Thus, the block set {·, I} and {+, I} as well as {·, +, I} are functionally complete. In a sense, the last set of blocks is more than complete, but, as we have seen, it is common practice to use this set so as not to restrict and complicate implementation of functions.

There are *single compound* blocks which are functionally complete in themselves and which have wide current use.

The first one we shall investigate is called NAND, and its logic block will be denoted by N·. It has any number of inputs and one output which is 1 only if at least *one* input is 0. Thus, for two variables (NAND):

A	B	f
0	0	1
0	1	1
1	0	1
1	1	0

We can write f as $A'B' + A'B + AB'$ and simplify it using a Karnaugh map:

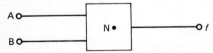

from which we read (simplified) f as $A' + B'$. Thus, the single NAND block diagram

A o——————
         N·————o f
B o——————

is equivalent to the (compound) block

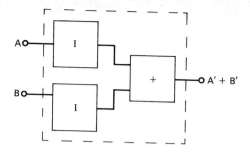

We now proceed to show that NAND is functionally complete. We plan to do this by proving that both AND and INVERT can be implemented using NANDs alone — since the set $\{\boxed{\cdot}, \boxed{I}\}$ is functionally complete. First,

it is clear that 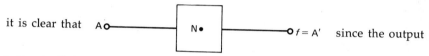 since the output

is 1 only if the input is 0. Now consider

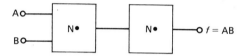

The output of the first NAND block is $A' + B'$ and that of the second block is the inversion of its input, namely, $(A' + B')'$ which reduces to AB. Therefore, NAND is functionally complete.

Incidentally, OR can also be implemented using NAND alone. Algebraically, $A + B = (A')' + (B')'$ so that

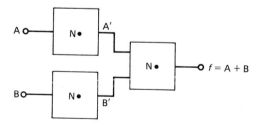

We can realize any Boolean function using NAND alone. For example, consider $f = A + BC'$. $A + BC' = (A')' + (B' + C)'$, and we should therefore NAND A' and $B' + C$. We know how to get A'. As for $B' + C$, it is $B' + (C')'$ and is thus formed by NANDing B and C'. We can easily construct C', so our plan is complete.   The diagram follows:

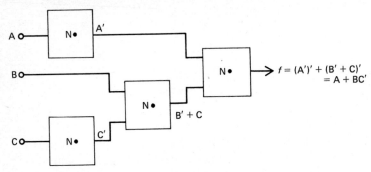

Another often-used compound block is NOR.   We shall use the block symbol $\boxed{N+}$ for NOR.   It has any number of inputs, and the output is 1 only if *all* its inputs are 0.   So, for two variables, we have

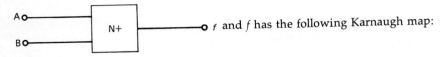 *f* and *f* has the following Karnaugh map:

Thus, $f = A'B'$, the dual of $A' + B'$.   Again we show that NOR is functionally complete by showing that INVERT and AND can be implemented using NOR alone.

First,

Then, since $AB = (A')'(B')'$,

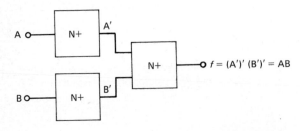

Thus, NOR is functionally complete. OR can be constructed using NOR alone. $A + B = (A'B')'$, so

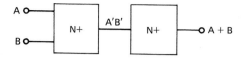

Any Boolean function can be implemented using NOR alone. Many times a judicious use of De Morgan's laws can reduce the number of NOR blocks necessary to implement a function. Consider $f = A'B + C$. We could construct this function directly by simply using developments for $+$ and $\cdot$ given above. We would then have

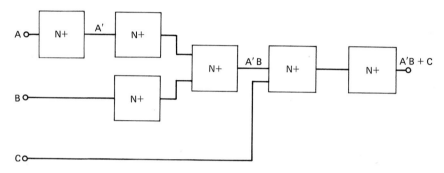

But now consider that $A'B + C$ can be written as $[(A'B)'C']'$ and this can be written $\{[A'(B')']'C'\}$. Thus we have the following circuit:

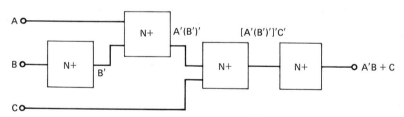

Other compound logic blocks are in current use. We shall introduce some of them in the exercises that follow, but we present just one more as this section ends. It is called EXCLUSIVE OR, and we use the logic block symbol $\boxed{E+}$. It has two inputs A and B, and its truth table is as follows:

	A	
B	0	1
0		1
1	1	

Thus, $f = AB' + A'B$. EXCLUSIVE OR is *not* functionally complete. Note that its output is 1 if either A or B is 1 but *not* if they are both 1.

8.22 Recall the EXCLUSIVE OR logic block $\boxed{E+}$ discussed above:

$A \circ$ — $\boxed{E+}$ — $\circ f = A \oplus B$. Use truth tables to prove (or disprove)
$B \circ$

that the following are identities:

(a) $A \oplus B = B \oplus A$
(b) $A(B \oplus C) = (AB) \oplus (AC)$
(c) $A + (B \oplus C) = (A + B) \oplus (A + C)$

8.23 Prove that the set $\{\boxed{E+}, \boxed{+}\}$ is functionally complete. Assume that constants 0 and 1 are available as inputs.

8.24 Draw a logic block diagram which realizes $f = A' + BC$

(a) using NAND blocks only;
(b) using NOR blocks only;
(c) using EXCLUSIVE OR and OR blocks only.

8.25 A logic block called MAJORITY which we denote by $\boxed{MAJ}$ has an *odd* number of inputs. The output is 1 only if more than half the inputs are 1 (that is, if the *majority* of its inputs are 1). For example, for three inputs A, B, and C, the output is 1 if two of the inputs are 1 or if all three are 1; it is 0 otherwise.

(a) Verify the output of the following:

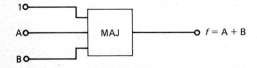

(b) Prove that the set $\{\boxed{MAJ}, \boxed{I}\}$ is functionally complete.
(c) Show that $f = AB$ can be constructed using $\boxed{MAJ}$ and $\boxed{I}$, assuming that 0 and 1 are available as inputs.

8.26 A function f has the following Karnaugh map:

| AB | | | | |
CD	0 0	0 1	1 1	1 0
0 0		1		
0 1		1		
1 1	1	1	1	1
1 0		1		1

(a) Draw the logic circuit which realizes f using the logic blocks $\boxed{\cdot}$, $\boxed{+}$ and $\boxed{I}$.

(b) Draw a logic circuit using $\boxed{\cdot}$ and $\boxed{I}$ only.

(c) Draw a logic circuit using $\boxed{N+}$ only.

8.27 A logic block called IMPLICATION denoted by $\boxed{\text{IMP}}$ has two inputs denoted by A and B. The output is 1 unless $A = 1$ and $B = 0$. We write algebraically $A \rightarrow B$.

(a) Place the values of $f = A \rightarrow B$ in a Karnaugh map and show that $f = A' + B$.

(b) Assuming that constants 0 and 1 are available as inputs, show that $\boxed{\text{IMP}}$ is functionally complete.

(c) Draw a logic circuit which realizes $f = A' + BC$ using $\boxed{\text{IMP}}$ only.

8.7 APPLICATIONS OF BOOLEAN ALGEBRA: HALF– AND FULL–ADDERS

We now turn to some of the ways Boolean algebra and logic block circuits are used to design circuits that perform arithmetic in the digital computer. Addition is the basic operation; binary numbers are used. Recall that the other operations can all be reduced to addition. We first reinterpret the values 0 and 1 that Boolean variables take on. Let us now consider that when a signal appears on a "wire" the binary digit or *bit* 1 is present at that input position, and when no signal appears on a wire the binary bit 0 is present.

We first construct what is called a *half-adder*. It is a logic circuit which actually adds two binary bits. Recall the arithmetic of adding binary digits:

$$
\begin{array}{cccc}
1 & 1 & 0 & 0 \\
+\ 1 & +\ 0 & +\ 1 & +\ 0 \\
\hline
10 & 01 & 01 & 00
\end{array}
$$

We have represented each sum as two digits. The rightmost digit of the sum is called the *sum digit*. We clearly have the following table for the sum digit of two digits A and B:

A B	0	1
0	0	1
1	1	0

From this table we read out the sum digit function $f_S = AB' + A'B$. We can implement f_S in various ways using the methods of Sec. 8.6. Any functionally complete set of logic blocks can be used. We note, incidentally, that $AB' + A'B$ recalls the logic block EXCLUSIVE OR, $\boxed{E+}$. However, we shall here revert to the more basic set $\{\boxed{+}, \boxed{\cdot}, \boxed{I}\}$. We thus have the following logic circuit for f_S:

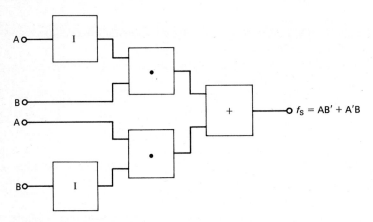

In this diagram we have repeated the inputs A and B rather than introduce "wires" that cross each other.

The lefthand digit in the sum above is called the *carry* digit. We can plot its values in a table:

A B	0	1
0		
1		1

from which we read the carry function $f_C = AB$. The simple logic circuit which realizes this function appears thus:

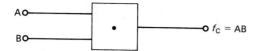

$f_C = AB$

We now place both of these circuits in one circuit:

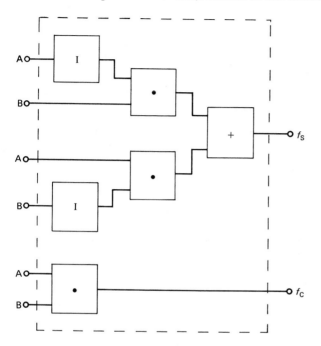

We now have the half-adder. This *compound* block is denoted thus:

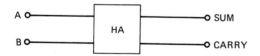

where *HA* stands for half-adder.

We have now designed a logic circuit which is capable of adding two binary bits.

Let us now consider the more general problem of adding two binary numbers, as in

```
  110111
+ 111001
```

We immediately note that in order to perform this addition we shall have to be able to add *three* binary bits—we have to carry over the carry digit to the next column each time. This will lead us to the design of the *full-adder*.

The full-adder has three inputs, and its output is once again a sum digit and a carry digit. The addition table for three binary bits A, B, and C leads to the following Karnaugh map for the *sum digit*:

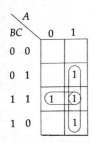

The map reveals that no simplification is possible. Thus, $f_S = AB'C' + A'B'C + ABC + A'BC'$.

As for the *carry digit*, we have the following Karnaugh map:

BC \ A	0	1
0 0		
0 1		1
1 1	1	1
1 0		1

Using the three 2-loops, we read out $f_C = AC + BC + AB$. The full-adder is denoted thus:

A ○——————┌──────┐——————○ f_S ($= AB'C' + A'B'C + ABC + A'BC'$)
B ○——————│ FA │
C ○——————└──────┘——————○ f_C ($= AC + BC + AB$)

We now want to construct the logic block diagram for the full-adder. We first present a circuit drawn directly from the function for f_S and f_C given above.

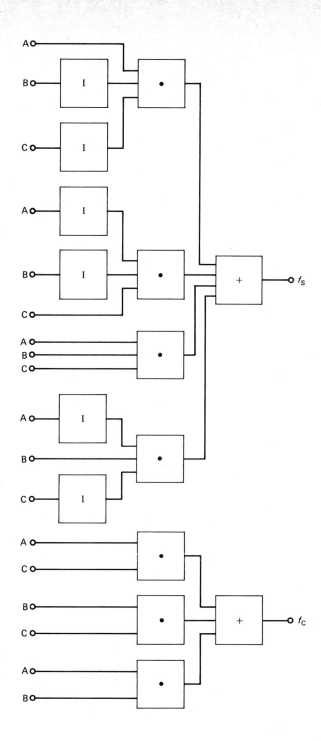

This circuit makes use of six inverters, seven ANDs, and two ORs. A somewhat simpler version of this circuit can be achieved by first doing a little analysis using Boolean algebra. We note first that $f'_c = A'B'C' + AB'C' + A'B'C + A'BC'$. This can be found by expanding f_c into sum normal form and then including in f'_c all those terms of the complete sum normal form in three variables which do not appear in f_c. Next, we note that

$$(A + B + C)f'_c = (A + B + C)(A'B'C' + AB'C' + A'B'C + A'BC')$$
$$= AB'C' + A'BC' + A'B'C$$

This last sum is precisely f_s except for the term ABC. In our new circuit we plan to first construct f_c, as we did above. Then we shall implement $A + B + C$ and f'_c and, making use of the algebra just shown, implement all f_s except for the term ABC. This term will be implemented separately and added in. This is the new circuit:

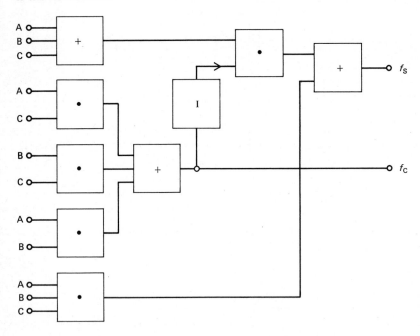

We have now used just one inverter, five ANDs, and three ORs, a total of six less of these blocks.

Our last project will be to put together a collection of adders to compose a circuit which will add two binary numbers. Suppose we take two four-digit binary numbers which we shall denote by $x_4x_3x_2x_1$ and $y_4y_3y_2y_1$, where each x_i and y_i is either 0 or 1.   Let us denote the sum of these two numbers by $s_5s_4s_3s_2s_1$.   Thus,

$$
\begin{array}{r}
x_4x_3x_2x_1 \\
+\ \ y_4y_3y_2y_1 \\
\hline
s_5\,s_4\,s_3\,s_2\,s_1
\end{array}
$$

Example

$$
\begin{array}{r}
1011 \\
+\ 1001 \\
\hline
10100
\end{array}
$$

There are many ways to construct such a circuit.   We shall show two ways. The first method uses just one full-adder over and over again.   An important facet of the process that we shall not be able to discuss is the *timing* and *sequencing* of the appearance of the various signals to the input of the adder. To make this first method work properly will evidently require an elaborate network of preliminary timing and sequencing circuits.   The proper allocation of the output of the adder will also be required so as to produce the correct sum.   Here is how this method works.   At a certain appropriate time t_1, the signals representing 0, x_1, and y_1 appear at the input terminals of the full-adder.   The sum digit output s_1 is transmitted away and saved.   At the next appropriate time t_2 (after a certain delay), the carry digit output c_1 is moved to the *input* of the adder and at the same time the next two digit signals x_2 and y_2 are inputted to the adder.   Once again the sum digit s_2 is transmitted and saved; once again the carry digit c_2 is fed into the adder together with the next two digits x_3 and y_3.   This occurs again, after a suitable delay, at time t_3.   The process is repeated until s_4 has been transmitted and saved.   At this time the carry digit c_4 is actually the sum digit s_5.   It can then be transmitted as s_5 and the sum $s_5s_4s_3s_2s_1$ displayed.   Another method would be to carry c_4 once again as input to the adder and cause 0 and 0 to also be fed in as input.   The output would then be s_5 with carry digit 0.   A pictorial view of this method follows. Note that only one *FA* is used.

The second scheme uses as many full-adders as there are digits in the sum.   All the digits of both of the numbers to be added are fed in to the set of full-adders at one time instead of sequentially.   The pictorial view we have just given can be used for this method, too.   The sequencing times t_i will not appear, however.   Instead, at the appropriate time t, preceding circuitry will cause *all* the x_i and y_i and the required 0s to appear at the designated input terminals.   The signals would then be held until the output signals s_i appear.

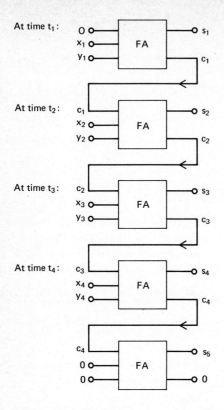

Different preliminary circuitry and five full-adders are used in this method. The careful sequencing of sets of digits we used in the previous method is eliminated but at the expense of supplying four additional full-adders.

EXERCISES

8.28 Design a full-adder using half-adders only.

8.29 Redesign the full-adder circuit using NAND only.

8.30 Think about how the circuit of full-adders (either version) which adds two four-digit binary numbers could be used to *subtract* two binary numbers. Recall the subtraction methods presented in Chap. 1.

ANSWERS TO SELECTED EXERCISES

Chapter 1

1.1 (a) 1.(45)   (c) 0.(571428)   (e) 8.75(0)

1.2 (a) $^{23}/_{99}$ (c) $-^{65}/_{1998}$ (e) $^{511}/_{1111}$ (g) $^{113}/_9$ (i) $^{707}/_{99900}$

1.4 (a) $6 \times 10^4 + 7 \times 10^3 + 9 \times 10^2 + 9 \times 10^1 + 5 \times 10^0$
     (c) $4 \times 10^1 + 5 \times 10^{-1} + 8 \times 10^{-2} + 6 \times 10^{-3}$ (e) $5 \times 10^2 + 9 \times 10^{-3}$

1.5 (a) 6808   (c) 75.04   (e) 406.778

1.6 (a) 0.486247×10^4 (c) 0.3370×10^9 (e) 0.9347×10^{20}
     (g) -0.4785003×10^4

1.7 (a) 0.86937 +03   (c) 0.625 +19   (e) 0.497 −03

1.8 (a) .486247E+04   (c) .3370E+09   (e) .9347E+20
     (g) −.4785003E+04

1.9 (a) 5   (c) 5   (e) 5   (g) 6   (i) 2

1.10 (a) 17.32   (c) 0.0033   (e) 985440, 985400, 985000, 990000
      (g) 789.1, 790.

1.11 (a) 17.32   (c) 0.0032   (e) 985430, 985400, 985000, 980000
      (g) 789.0, 780.

1.12 (a)   17.3 14.1   (c)   8.267   692.   (e)   138.750 70.7
             .82   5.92   − 67.983
             ───   13.2   ───────
             346   17.   70.767
             1384   647.8957
             ─────   692.2827
             14.186

(g) $\begin{array}{r} 329 \\ 3\overline{)987} \\ 329. \end{array}$ 335. (i) $\begin{array}{r} 89.5 \\ 46.7 \\ \hline 136.2 \end{array}$ $\begin{array}{r} 10.47692 \ldots \\ 13\overline{)136.2} \end{array}$

$\quad\quad\quad \dfrac{6.389}{335.389}$ $\quad\quad\quad\quad\quad\quad\quad \begin{array}{r} 10.47692 \ldots \quad 0.906 \\ -\ 9.57 \\ \hline 0.90692 \ldots \end{array}$

1.13 (a) $\begin{array}{r} 64280. \\ 23.75 \\ \hline 64303.75 \end{array}$ $\begin{array}{r} 64300. \\ -\ 575 \\ \hline 63725 \end{array}$ | 6 | 3 | 7 | 2 | + | 0 | 0 | 5 |

(c) $\begin{array}{r} 1.181 \ldots \\ 7\overline{)8.27} \end{array}$ $\begin{array}{r} 1.181 \\ -\ 0.3728 \\ \hline 0.8082 \end{array}$ | 8 | 0 | 8 | 2 | + | 0 | 0 | 0 |

(e) $\begin{array}{r} 8.37 \\ 0.6578 \\ \hline 9.0278 \end{array}$ $\begin{array}{r} 9.027 \\ 17.379 \\ \hline 26.406 \end{array}$ $\begin{array}{r} 26.40 \\ 4.72 \\ \hline 5280 \\ 18480 \\ 10560 \\ \hline 124.6080 \end{array}$ | 1 | 2 | 4 | 6 | + | 0 | 0 | 3 |

(g) $\begin{array}{r} 56.78 \\ 68.79 \\ \hline 125.57 \end{array}$ $\begin{array}{r} 125.5 \\ 45.68 \\ \hline 171.18 \end{array}$ $\begin{array}{r} 171.1 \\ -\ 34.77 \\ \hline 136.33 \end{array}$ $\begin{array}{r} 34.075 \\ 4\overline{)136.3} \end{array}$

| 3 | 4 | 0 | 7 | + | 0 | 0 | 2 |

1.14 (a) | 6 | 3 | 7 | + | 0 | 5 | (c) | 8 | 0 | 7 | + | 0 | 0 |

(e) | 1 | 2 | 4 | + | 0 | 3 | (g) | 3 | 3 | 7 | + | 0 | 2 |

1.15 (a) | 6 | 4 | 7 | 2 | 8 | + | 0 | 5 | (c) | 8 | 0 | 8 | 6 | 0 | + | 0 | 0 |

(e) | 1 | 2 | 4 | 6 | 3 | + | 0 | 3 | (g) | 3 | 4 | 1 | 2 | 0 | + | 0 | 2 |

Chapter 2

2.1 (a) 989   (c) 10995   (e) 425   (g) 194   (i) 3173

2.2 (a) $^{29}/_{49} = 0.(591836734693879) \doteq 0.59183$

(c) $^{1959}/_{3125} = 0.62688(0) = 0.62688$

(e) $^{2999}/_{4096} = 0.732177 \ldots \doteq 0.73217$

(g) $^{91}/_{243} = 0.374485 \ldots \doteq 0.37448$

(i) $^{13}/_{128} = 0.1015625(0) \doteq 0.10156$

2.3 (a) 11.(5)   (c) 19.488(0)   (e) 123.60351 . . .   (g) 136.5(0)

(i) 1845.69822 . . .

2.5 (a) $1001111011_{(2)}$, $21323_{(4)}$, $10020_{(5)}$ (c) $11011120_{(3)}$, $110110_{(6)}$, $23346_{(9)}$

(e) $10000221100_{(3)}$, $355523_{(7)}$, $42467_{(11)}$

2.6 $10212_{(5)}$

2.7 (a) $15302_{(6)}$, $212312_{(4)}$, $4666_{(8)}$

  (c) $121120110_{(3)}$, $14244_{(5)}$, $2257_{(8)}$

  (e) $100100_{(8)}$, $11000001000000_{(2)}$, $133104_{(6)}$

2.8 (a) $0.00001(0110)_{(2)}$, $0.002(30)_{(4)}$, $0.02(6314)_{(8)}$

  (c) $0.235404335 \ldots _{(6)}$, $0.210(14)_{(5)}$, $0.130112 \ldots _{(4)}$

  (e) $0.00(03)_{(5)}$, $0.002436 \ldots _{(8)}$, $0.00(1434)_{(7)}$

2.9 (a) $1110101.00001_{(2)}$, $100002.00112_{(3)}$, $3311.00331_{(4)}$

  (c) $50A4.67249_{(12)}$, $3CB2.73333_{(13)}$, $322C.79B42_{(14)}$

  (e) $100.02211_{(4)}$, $20.12257_{(8)}$, $10.285E9_{(16)}$

2.10 (a) $3123_{(4)}$, $333_{(8)}$, $DB_{(16)}$ (c) $110331.31_{(4)}$, $2475.64_{(8)}$, $53D.D_{(16)}$

2.11 (a) $11101011011000_{(2)}$ (c) $0.111111100110110_{(2)}$

  (e) $111001.110111100111_{(2)}$

2.12 $B93_{(16)}$

2.14 $5378_{(9)}$

2.15 (a) $14030_{(6)}$ (c) $1268960_{(12)}$ (e) $11101011010_{(2)}$

  (h) $25.14362_{(7)}$

2.16 (a) $11022020_{(3)}$ (c) $341215_{(6)}$ (e) $47804_{(14)}$

2.17 (a) 1101101   (c) 4023561   (e) 1234567

      0101101   4060031   0543210

   1 0011010   1 1113622   1 777777

         1   1   1

      $11011_{(2)}$ $1113623_{(7)}$ $777778_{(16)}$

  (g) 3030303

      0032110

      3123013

    $- 210320_{(4)}$

2.18 (a) 13013   (c) 4A3   (e) 4775   (g) 4432

        231   9A   1247   302

      13013   4538   42653   14414

      111111   4045   24764   24401

      32032 $44988_{(11)}$ 11772 $3010014_{(5)}$

    $10333323_{(4)}$ 4775

                               $6506713_{(8)}$

2.19 (a) $332.2_{(5)}$ (c) $1256.1_{(7)}$ (e) $10112.3_{(4)}$

2.23 (a) Decimal   Binary coded

        7   0111

        5   0101

       12   1100 #

                     0110

              0001   0010

                1   2

(c) Decimal   Binary coded

Decimal	Binary coded	
7 5	0111	0101
4 8	0100	1000
1 2 3	1011#	1101#
	0110	0110
0001 0010	0010	0011
1 2		3

(e) Decimal

Decimal	Binary coded			
1 4 7 3	0001	0100	0111	0011
9 2 4 3	1001	0010	0100	0011
1 0 7 1 6	1010#	0110	1011#	0110
	0110		0110	
0001 0000	0000	0111	0001	0110
1 0		7	1	6

Chapter 3

3.1 (a) 14   (c) -14 (e) $^{13}/_4$ (g) -3

3.2 (a) $x = 0$ (c) $a = 4$ (e) $x = 21$

3.3 $p + q = -8$ (c) $4x + 3y = 15$ (e) $a^2 = 37$ (g) $d = vt$

3.4 (a) $(8,0)$, $(1,^7/_3)$, $(0,^8/_3)$, $(3,^5/_3)$, $(2,2)$
     (c) $8 + 28 - 16 - 67 = -47 \neq 0$

3.5 (a) $3x - 4 = 131$ (c) $10x + 201 = 2010 - 13x$
           $3x = 140$ $23x = 1203$
            $x = 30_{(5)}$ $x = 21_{(4)}$

3.6 (a) $^{19}/_8$ (c) $^{21}/_8$ (e) $^{15}/_2$ (g) $^{15}/_8$ (i) $-^1/_4$ (k) 64

3.7 $q = \dfrac{3x + 4y - 8}{5}$

3.8 $y = x^2 - 6$, $x = \pm(6 + y)^{1/2}$

3.10 $f(x + h) - f(x) = 2xh + h^2$

3.11 $g(5) = ^1/_{10}$; $g(s - 5) = \dfrac{s - 9}{s}$; $\dfrac{4}{g(r)} = \dfrac{4(5 + r)}{r - 4}$

3.12 $h(1,2) = -6$; $h\left(\dfrac{1}{a}, \dfrac{1}{b}\right) = \dfrac{1}{a} - \dfrac{4}{b} + a$; $h(3,0) = ^{10}/_3$

3.13 $G(0,0,2) = 4$; $G(1,0,s^{1/2}) = 1 + s$; $G(-1,-1,-3) = 13$

3.14 (a) $x + 5y = 23$ (c) $y = 1$ (e) $x = 5$

3.16 (a) $m = 1$ (c) $m = 1$ (e) $m = 0$

3.17 (a) $y = \dfrac{3x}{2}$ (c) $y = 6x + 4$ (e) $5x - 3y + 9 = 0$

3.19 $F = \dfrac{9C}{5} + 32$

3.20 (b) $-40°$

3.21 (a) $s = 29$

3.22 (a) $x = {}^{29}/_{10}$, $y = {}^{17}/_{10}$ (c) $a = 10$, $b = -7$

  (e) $x = -1$, $y = -8$, $z = -2$ (g) $x = -{}^{488}/_{21}$, $y = {}^{100}/_3$

3.23 (a) $x = 11_{(2)}$, $y = -1_{(2)}$ (c) $x \doteq 5.51_{(9)}$, $y \doteq 0.27_{(9)}$

3.24 (a)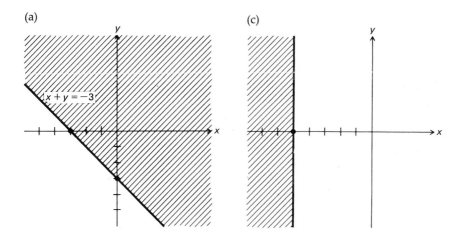

  (c)

  (f)

3.25 (a) $[4,\infty)$ (c) $(-\infty,3]$ (e) $[3,\infty)$

3.26 (a) $x = 4$ or $x = -6$ (c) $x = {}^{7}/_2$

3.27 (a) $y > {}^{16}/_3$ (c) $x \geqslant -8$ (e) $-15 < q < -10$

  (g) ${}^{13}/_6 < x < {}^{43}/_6$ (i) $8 \leqslant w < 16$

3.28 (a) x $-2 < x < 5$

  (c) x $2 \leqslant x < 7$

3.29 (a) $-3 < x < 3$ (c) $-{}^{36}/_5 < x < {}^{44}/_5$ (e) $x < 1$ or $x > {}^{11}/_3$

3.30

(a)

(c)

$x + y = -3$

(e)

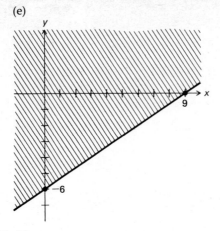

3.31 (a)

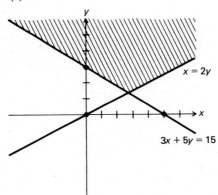

$x = 2y$

$3x + 5y = 15$

3.32 (a)

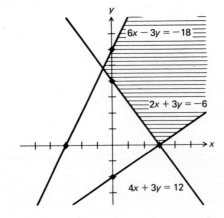

$6x - 3y = -18$

$2x + 3y = -6$

$4x + 3y = 12$

(c)

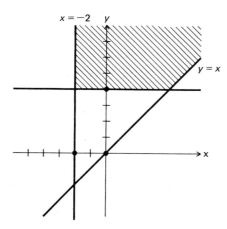

(e)

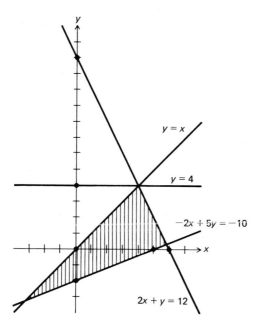

3.33 $z_{max} \doteq 51.74$ occurs at $x = {}^{110}/_{31}$, $y = {}^{72}/_{31}$

3.34 $z_{max} \doteq 494.2$ occurs at $(7, {}^{65}/_6)$

3.35

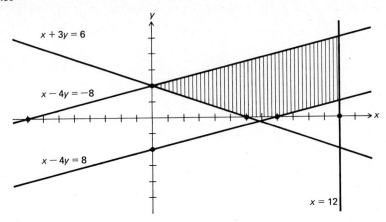

(a) $z_{min} = 11$ at $(0,2)$
(b) $z_{max} = 25$ at $(12,1)$

3.38

x	y	x²	xy
−1	2	1	−2
0	1	0	0
2	−5	4	−10
3	−8	9	−24
Σ 4	−10	14	−36

$$4a + 4b = -10$$
$$4a + 14b = -36$$
$$\overline{10b = -26}$$
$$b = -2.6$$
$$a = 0.4$$

$y = 0.4 - 2.6x$

3.40

x	y	x²	xy
1	−2	1	−2
2	−1	4	−2
3	−1	9	−3
4	0	16	0
5	1	25	5
6	3	36	18
7	5	49	35
Σ 28	5	140	51

$$7a + 28b = 5$$
$$28a + 140b = 51$$
$$\overline{28b = 31}$$
$$b = {}^{31}/_{28}$$
$$a = -{}^{26}/_{7}$$

$y = -{}^{26}/_{7} + \dfrac{31x}{28}$

Chapter 4

4.1 (a) integer   (c) real   (e) real   (g) integer
4.2 (a) .678E+03   (c) .935E−05   (e) .56784E+16
4.3 (a) real   (c) real   (e) integer   (g) real   (h) integer
    (k) not acceptable (contains first character which is not alphabetic)

(m) not acceptable (contains nonalphanumeric character)

(o) not acceptable (first character not alphabetic, contains nonalphanumeric character)

4.4 (a) $A - (B + C)/G$ (c) $A + (B - C)**.8$ (e) $A + B/C + D/E$

(g) $((X + 3)/N + K)**J)/5$ (i) $3470/D**2 + 7856/(17*R)$

(k) $1. + X + X**2 + X**3 + X**4$

4.5 (a) $(R/S)**(K+7)$ (c) $(X + 7)/(Y - 6)$ (f) $A/C + C*D/A$

(g) $6/(X*Y*Z)$

4.6 (a) integer   (c) mixed   (e) real

4.7 (a) 15.   (c) 0   (e) 16.   (g) 10.

4.8 (a) $Y = (2*X + 7) **.5$ (c) $V = 3.14159 * R **2$

(e) $Q = H*(V**7 + 3*W**6)/6$ (g) $R = (2 + (2 + S)**.5)**.5$

(i) $ALPHA = (4.*(BETA**2 - 4.)**1.5)/BETA$

4.9 (a) $A = 281.43$, $B = 8.267E+02$, $N = -25$, $X = 143.271$, $Y = 2.13997$,
$M = 41$

(c) $A = 28143.8$, $B = .267E+02$, $N = -2$, $X = 50143.27$, $Y = 1.2139$,
$M = 9741$

4.10 (a) 1 card, 8 variables   (c) 8 cards, 1 variable per card

(e) 2 cards, 4 variables per card

4.11 $X = 4.72$

$Y = -12.73$

$M = 7$

$N = -5$

$Z = X**2 + Y**(M+N)$

$Q = Z**.5 + (X**M + Y**N)/17.$

WRITE (3, 1) Z, Q

1   FORMAT (1X, 2E16.6)

END

4.14   READ (2, 1) A, B, C

1   FORMAT (F9.0)

$S = (A + B + C)/2.$

$AREA = (S*(S - A)*(S - B)*(S - C))**.5$

WRITE (3, 2) A, B, C

2   FORMAT (1X, 3E16.6)

WRITE (3, 3) S, AREA

3   FORMAT (1X, 2E16.6)

END

4.17 7 READ (2, 1) A, B, C, X

1   FORMAT (4F6.0)

IF $(C - (A + 3.*B))$ 2, 3, 4

2   $Y = X**5$

GO TO 5

```
    3   Y = X**4
        GO TO 5
    4   Y = X**3
    5   WRITE (3, 6) Y
    6   FORMAT (1X, E 13.6)
        GO TO 7
        END
4.18    READ (2, 1) A, B, C, D
    1   FORMAT (4F10.0)
        IF (A − B) 2, 2, 3
    2   BIG = B
        GO TO 9
    3   BIG = A
    9   IF (BIG − C) 44, 44, 5
   44   BIG = C
    5   IF (BIG − D) 6, 6, 7
        BIG = D
    7   WRITE (3, 8) BIG
    8   FORMAT (1X, E13.6)
        END
4.21    READ (2, 1) N, R, S
    1   FORMAT (I4, 2F1010)
        IF (N − 1) 3, 2, 3
    2   IF (R − S) 4, 5, 5
    4   Z = 4.*R − 3.*S
        GO TO 6
    5   Z = 3.*S − 4.*R
    3   I = 567
        WRITE (3, 7) I
    7   FORMAT (1X, I5)
        GO TO 8
    6   WRITE (4, 9) Z
    9   FORMAT (1X, E13.6)
    8   CONTINUE
        END
```

Chapter 5

```
5.2     DIMENSION A(40)
        I = 1
    2   READ (2, 1) A(I)
    1   FORMAT (F6.0)
```

```
      I = I + 1
      IF (I − 40) 2, 3, 3
   3  READ (3, 4) M
   4  FORMAT (I2)
      IF (M − 1) 5, 6, 6
   6  IF (M − 40) 7, 7, 5
   7  I = 1
      SUM = 0
  10  SUM = SUM + A(I)
      IF (I − M) 8, 11, 11
   8  I = I + 1
      GO TO 10
  11  WRITE (3, 12) SUM
  12  FORMAT (1X, E13.6)
      GO TO 13
   5  WRITE (3, 9) M
   9  FORMAT (1X, I4)
  13  CONTINUE
      END
5.4   DIMENSION A(20), B(30)
      I = 1
   4  READ (2, 1) A(I), A(I+1)
   1  FORMAT (2F10.0)
      IF (I − 17) 2, 2, 3
   2  I = I + 2
      GO TO 4
   3  I = 1
   8  READ (2, 5) B(I)
      I = I + 1
      IF (I − 29) 6, 6, 7
   6  I = I + 1
      GO TO 8
   7  I = 1
      SUM = 0
  11  SUM = SUM + A(I)
      IF (I − 19) 9, 9, 10
   9  I = I + 1
      GO TO 11
  10  I = 1
      TOT = 0
  14  TOT = TOT + B(I)
      IF (I − 29) 12, 12, 13
```

```
      12   I = I + 1
           GO TO 14
      13   S = SUM + TOT
           WRITE (3, 15) S
      15   FORMAT (1X, E13.6)
           END
5.6        DIMENSION G(40)
           I = 1
       5   READ (2, 1) G(I), G(I+1), G(I+2), G(I+3)
       1   FORMAT (4F10.0)
           IF (I − 33) 3, 3, 4
       3   I = I + 4
           GO TO 5
       4   I = 1
           SUM = 0
       8   SUM = SUM + G(I)**2
           IF (I − 39) 6, 6, 7
       6   I = I + 1
           GO TO 8
       7   X = SQRT(SUM)
           WRITE (3, 9) X
       9   FORMAT (1X, E13.6)
           END
```

5.7 (a) $\mathbf{C} = \begin{bmatrix} 33 & 24 \\ 29 & 28 \end{bmatrix}$, $\mathbf{D} = \begin{bmatrix} -1 & 2 \\ 25 & 52 \end{bmatrix}$ (c) $\mathbf{A}^2 = \begin{bmatrix} 21 & 48 \\ 16 & 37 \end{bmatrix}$,

$\mathbf{B}^2 = \begin{bmatrix} 37 & -4 \\ -7 & 32 \end{bmatrix}$, $\mathbf{A}^2 - \mathbf{B}^2 = \begin{bmatrix} -16 & 52 \\ 23 & 5 \end{bmatrix}$ (e) $5\mathbf{A}^2 = \begin{bmatrix} 105 & 240 \\ 80 & 185 \end{bmatrix}$,

$4\mathbf{AB} = \begin{bmatrix} 132 & 96 \\ 116 & 112 \end{bmatrix}$, $\mathbf{B}^2 = \begin{bmatrix} 37 & -4 \\ -7 & 32 \end{bmatrix}$, $5\mathbf{A}^2 + 4\mathbf{AB} + \mathbf{B}^2 = \begin{bmatrix} 274 & 332 \\ 189 & 329 \end{bmatrix}$

5.8 $\mathbf{D} \times \mathbf{C} \times \mathbf{A} = \begin{bmatrix} 90 \\ 150 \end{bmatrix}$, $\mathbf{B} \times \mathbf{E} = \begin{bmatrix} 28 & 45 \end{bmatrix}$

5.9 $\begin{bmatrix} 5 & 4 \\ 2 & 7 \end{bmatrix} \begin{bmatrix} x \\ y \end{bmatrix} = \begin{bmatrix} 7 \\ 15 \end{bmatrix}$, $\mathbf{A}^{-1} = \frac{1}{27} \begin{bmatrix} 7 & -4 \\ -2 & 5 \end{bmatrix}$, $\begin{bmatrix} x \\ y \end{bmatrix} = \frac{1}{27} \begin{bmatrix} 7 & -4 \\ -2 & 5 \end{bmatrix} \begin{bmatrix} 7 \\ 15 \end{bmatrix} =$

$\frac{1}{27} \begin{bmatrix} -11 \\ 61 \end{bmatrix}$, $x = -11/27$, $y = 61/27$

5.11 $\mathbf{A}^{-1} = \frac{1}{76} \begin{bmatrix} -8 & -4 & 18 \\ 18 & -10 & 7 \\ -2 & 18 & -5 \end{bmatrix}$

5.13 (a) $\begin{bmatrix} 1 & 2 & 1 & 11 \\ 2 & -1 & -1 & -5 \\ -2 & 2 & 5 & 7 \end{bmatrix} \approx \begin{bmatrix} 1 & 2 & 1 & 11 \\ 0 & -5 & -3 & -27 \\ 0 & 6 & 7 & 29 \end{bmatrix} \approx$

$$\begin{bmatrix} 1 & 2 & 1 & 11 \\ 0 & -5 & -3 & -27 \\ 0 & 0 & 4 & -4 \end{bmatrix} \approx \begin{bmatrix} 1 & 2 & 1 & 11 \\ 0 & -5 & -3 & -27 \\ 0 & 0 & 1 & -1 \end{bmatrix} \approx \begin{bmatrix} 1 & 2 & 1 & 11 \\ 0 & 1 & 0 & 6 \\ 0 & 0 & 1 & -1 \end{bmatrix} \approx$$

$$\begin{bmatrix} 1 & 0 & 1 & -1 \\ 0 & 1 & 0 & 6 \\ 0 & 0 & 1 & -1 \end{bmatrix} \approx \begin{bmatrix} 1 & 0 & 0 & 0 \\ 0 & 1 & 0 & 6 \\ 0 & 0 & 1 & -1 \end{bmatrix}. \quad \text{Thus, } x = 0, y = 6, z = -1.$$

5.14 (a) 33   (c) 0

5.15 (a) 90   (c) -0.116 (e) 0

5.16 (a) $\det A = -103$

5.17 (a) $\mathbf{A} + \mathbf{B} = \begin{bmatrix} 6 & 11 \\ 2 & 4 \end{bmatrix}$, $\det(A + B) = 2$

$\mathbf{A} \times \mathbf{B} = \begin{bmatrix} -7 & 59 \\ 19 & 9 \end{bmatrix}$, $\det(A \times B) = -1184$

$\det A = -37$, $\det B = 32$, $\det A + \det B = -5$, $\det A \times \det B = -1184$

5.18 (a) $x = \dfrac{\begin{vmatrix} 56 & 5 \\ -3 & -7 \end{vmatrix}}{\begin{vmatrix} 3 & 5 \\ 2 & -7 \end{vmatrix}} = \dfrac{377}{11}$, $y = \dfrac{\begin{vmatrix} 3 & 56 \\ 2 & -3 \end{vmatrix}}{-31} = \dfrac{121}{31}$

5.19 (a) $x = \dfrac{\begin{vmatrix} 7 & 5 \\ 11 & 10 \end{vmatrix}}{0}$, $y = \dfrac{\begin{vmatrix} 3 & 7 \\ 6 & 11 \end{vmatrix}}{0}$ (b) $(0,-7)$, $(4,1)$, $(\frac{7}{2},0)$, $(3,-1)$

5.21 Let $\mathbf{A} = \begin{bmatrix} a & b \\ c & d \end{bmatrix}$. Then $\mathbf{A}^{-1} = \dfrac{1}{ad - bc} \begin{bmatrix} d & -b \\ -c & a \end{bmatrix} =$

$\begin{bmatrix} \dfrac{d}{ad - bc} & \dfrac{-b}{ad - bc} \\ \dfrac{-c}{ad - bc} & \dfrac{a}{ad - bc} \end{bmatrix}$. Thus, $\det A^{-1} = \dfrac{ad - bc}{(ad - bc)^2} = \dfrac{1}{ad - bc} = \dfrac{1}{\det A}$.

Chapter 6

6.1 (a) $-10i$ (c) $32i$ (e) $\dfrac{18}{13} - \dfrac{14}{13}i$ (g) $-i$

6.2 (a) $x^2 - 4x + 3 = (x - 3)(x - 1) = 0$; $x = 1$, $x = 3$
   (c) $10x^2 - 11x - 6 = (5x + 2)(2x - 3) = 0$; $x = -\frac{2}{5}$, $x = \frac{3}{2}$
   (e) $9y^2 - 52y + 35 = (9y - 7)(y - 5) = 0$; $y = \frac{7}{9}$, $y = 5$

6.3 (a) $x = 4 \pm 2\sqrt{6}i$ (c) $\frac{1}{2}, \frac{1}{2}$ (e) $-\frac{7}{6} \pm \dfrac{\sqrt{95}}{6}i$

   (g) $-2 \pm \sqrt{2}i$ (i) $q = \pm \dfrac{3}{2}i$

6.4 (a) $x = -\frac{1}{3}$, $x = -1$ (c) $x = \frac{35}{12}$, $x = \frac{7}{2}$ (e) $x = 27$, $x = 1$

6.5 (a) $y - 25 = 7(x^2 - 4x) = 7(x^2 - 4x + 4) - 28$. Thus, $y + 3 = 7(x - 2)^2$. $V(2,-3)$; upward; y intercept 25; x intercepts $2 \pm \sqrt{\dfrac{3}{7}}$.

(c) $128y = 63 - 16(x^2 - \frac{1}{2}x) = 63 - 16(x^2 - \frac{1}{2}x + \frac{1}{16}) + 1$; $128y - 64 = -16(x^2 - \frac{1}{2}x + \frac{1}{16}) = -16(x - \frac{1}{4})^2$; thus, $y - \frac{1}{2} = -\frac{1}{8}(x - \frac{1}{4})^2$; $V(\frac{1}{4}, \frac{1}{2})$; downward; x intercepts $\frac{9}{4}$ and $-\frac{7}{4}$; y intercept $\frac{63}{128}$.

(e) $y + 1 = 8x - 2x^2 = -2(x^2 - 4x) = -2(x^2 - 4x + 4) + 8$; thus, $y - 7 = -2(x - 2)^2$; $V(2,7)$; downward; x intercepts $2 \pm \sqrt{\dfrac{7}{2}}$; y intercept -1.

6.9 (c) $y = \dfrac{9}{14}x^2 + \dfrac{57}{70}x - \dfrac{19}{35}$

6.11 (c) $f(x) = x^3 - 87$, $f(4) = -23$, $f(5) = 38$
$f(4.5) = (4.5)^3 - 87 = 91.125 - 87 = 4.125$
$f(4.75) \doteq 20.17$
$f(4.37) \doteq -3.55$
$f(4.43) \doteq -0.062$
$f(4.46) \doteq 1.717$
$f(4.45) \doteq 1.121$
$f(4.44) \doteq 0.528$
Thus, $x \doteq 4.43$

6.12 (a) $x^3 - 3x - 15 = 0$. $f(x) = x^3 - 3x - 15$, $f'(x) = 3x^2 - 3$. The right side of Newton's formula is, therefore,

$$x - \frac{x^3 - 3x - 15}{3x^2 - 3} = \frac{2x^3 + 15}{3(x^2 - 1)}$$

The iterative formula is

$$x_{i+1} = \frac{2x_i^3 + 15}{3(x_i^2 - 1)}$$
$$x_1 = 3$$

We then have

$$x_2 = \frac{54 + 15}{3(8)}$$
$$\doteq 2.87$$

and

$$x_3 = \frac{2(2.87)^3 + 15}{3(2.87^2 - 1)}$$
$$\doteq 2.869$$

so that $x \doteq 2.869$.

6.14 $x_{i+1} = \sqrt{\dfrac{7}{x_i^2 + 3}}$, $x_1 = 1$. Thus, $x_2 = \sqrt{\dfrac{7}{4}} \doteq 1.323$, $x_3 = \sqrt{\dfrac{7}{1.75 + 3}} \doteq$

1.22, $x_4 = \sqrt{\dfrac{7}{4.475}} \doteq 1.25$, $x_5 = \sqrt{\dfrac{7}{4.56}} \doteq 1.24$, $x_6 = \sqrt{\dfrac{7}{4.53}} \doteq 1.241$.

Thus, $x = 1.241$.   Check: $x^4 + 3x^2 - 7 = 0$, $x^2 = \dfrac{-3 + \sqrt{37}}{2} \doteq 1.5413$,

so that $x \doteq 1.24$.

6.15   X = 1.
    I = 1
1 X = (7./(X**2 + 3.))**.5
    I = I + 1
    IF (I − 10) 1, 1, 2
2 WRITE (3, 3) X
3 FORMAT (1X, E13.6)
    END

Chapter 7

7.1 (a) $3^y = 27$, $y = 3$.   (c) $10^y = 0.001$, $y = -3$.   (e) $9^y = 3$, $y = 0.5$
7.2 (a) $a^3 = 8$, $a = 2$ (c) $a^{3/2} = 27$, $a = 9$.
7.3 $e^{-0.18h} = 0.5$, $-0.18h = \ln 0.5 \doteq -0.693$, $h \doteq 3.85$ miles.
7.4 $i = 5e^{-3.8(1.5)} = 5e^{-5.7} \doteq 5(0.00335) = 0.01675$ amp.
7.6   X = 2.5
    I = 1
5 WRITE (3, 8) I, X
8 FORMAT (1X, I4, E16.6)
    X = (20. − EXP(X))/3.
    I = I + 1
    IF (I − 20) 5, 5, 6
6 CONTINUE
    END
7.8   X = 2.5
    I = 1
5 X = X − (EXP(X) + 3.*X − 20.)/(EXP(X) + 3.)
    I = I + 1
    IF (I − 10) 5, 5, 6

```
6   WRITE (3, 1)
1   FORMAT (1X, E13.6)
    END
```
7.11 $f(0.1) = 1.005$
7.12 $f(0.01) = 0.01000$
7.13 $f(0.1) = 1.1052$

Chapter 8

8.1 (a)

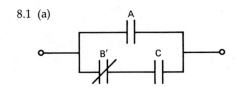

(c)

(e)

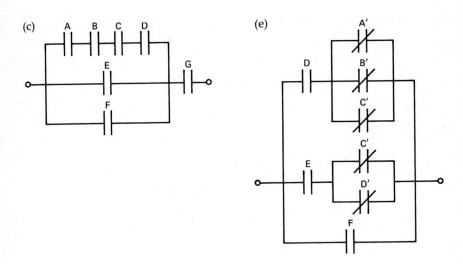

8.2 (a)

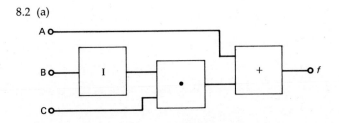

(c)

(e)

8.3 (a)

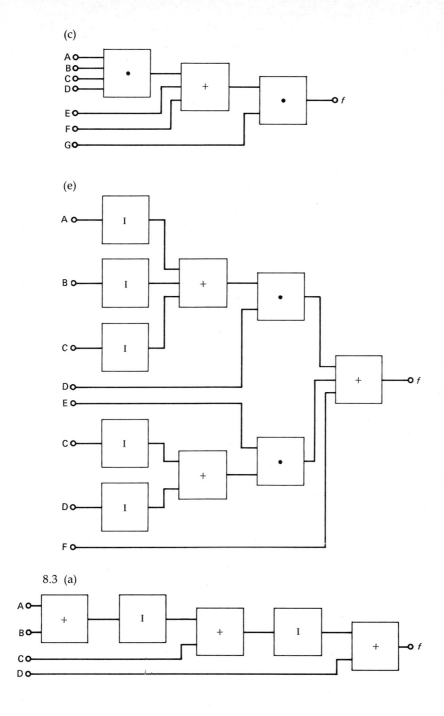

(c)

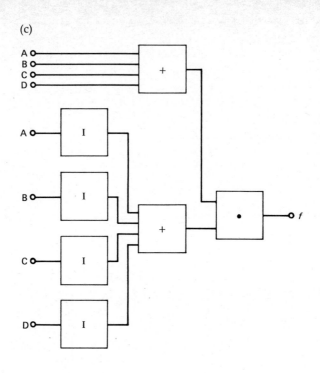

(e)

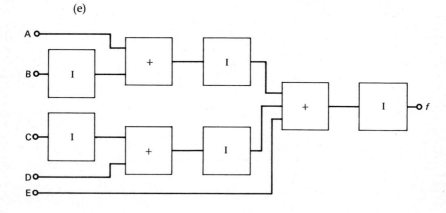

8.4 (a) $f = B(A'BC + B' + A')(A + B)$

(c) $f = AE' + (A + B)(C + D)E$

8.6 (a) $A(A' + AB) = AA' + AAB = 0 + AB = AB$

(c) $C(A + B'C')' = C[A'(B + C)] = A'BC + A'C = A'C$

(e) $(AB' + A'B)' = (AB')'(A'B)' = (A' + B)(A + B') = A'B' + AB$

(g) $A + B + C + A'B = B + C + (A + A')(A + B) = B + C + A + B = A + B + C$

(i) $(AB)' + (A'BC)' + A = A' + B' + A + B' + C' + A = 1$

(k) $B + AC + AB + C = B(1 + A) + C(1 + A) = B + C$

8.7 (a)

A	B	A + B	A'	A' + B	(A + B)(A' + B)
1	1	1	0	1	1
1	0	1	0	0	0
0	1	1	1	1	1
0	0	0	1	1	0

(c)

A	B	C	A + B + C	A(A + B + C)
0	0	0	0	0
1	0	0	1	1
1	0	1	1	1
1	1	0	1	1
1	1	1	1	1
0	1	1	1	0
0	1	0	1	0
0	0	1	1	0

(e)

A	B	AB	A'	A' + AB	(A' + AB)'	B'	AB'
1	1	1	0	1	0	0	0
1	0	0	0	0	1	1	1
0	1	0	1	1	0	0	0
0	0	0	1	1	0	1	0

(g)

A	B	C	A + B + C	A'	B'	A' + B'	A' + B' + C	(A + B + C)(A' + B' + C)	AB	A'B	AB' + A'B	A'B + A'B + C
0	0	0	0	1	1	1	1	0	0	0	0	0
1	0	0	1	0	1	1	1	1	1	0	1	1
1	0	1	1	0	1	1	1	1	1	0	1	1
1	1	0	1	0	0	0	0	0	0	0	0	0
1	1	1	1	0	0	0	1	1	0	0	0	1
0	1	1	1	1	0	1	1	1	0	1	1	1
0	1	0	1	1	0	1	1	1	0	1	1	1
0	0	1	1	1	1	1	1	1	0	0	0	1

8.8 (a) $A = A \cdot 1 = A(B + B') = AB + AB'$

(c) $A + B' = A(B + B') + B'(A + A') = AB + AB' + AB' + A'B' = AB + AB' + A'B'$

8.9 (a) $A = (AB + AB')(C + C') = ABC + AB'C + ABC' + AB'C'$

(c) $A + B' = ABC + AB'C + A'B'C + ABC' + AB'C' + A'B'C'$

8.10 (a) $B + C' = (A + A')B(C + C') + C'(A + A')(B + B') = ABC + ABC' + A'BC + A'BC' + AB'C' + A'B'C'$

(c) $A'B + B'C = A'B(C + C') + B'C(A + A') = A'BC + A'BC' + AB'C + A'B'C$

8.11 (a) $BC' + D = BC'(D + D')(A + A') + (A + A')(B + B')(C + C')D = BC'(AD + AD' + A'D + A'D') + (AB + AB' + A'B + A'B')(CD + C'D) = ABC'D + ABC'D' + A'BC'D + A'BC'D' + ABCD + AB'CD + A'BCD + A'B'CD + AB'C'D + A'B'C'D$

8.12 (a)

B \ A	0	1
0		1
1		1

(c)

B \ A	0	1
0		1
1	1	

8.13 (a)

BC \ A	0	1
0 0		
0 1	1	
1 1	1	1
1 0	1	1

(c)

BC \ A	0	1
0 0	1	
0 1		
1 1		
1 0	1	1

(e)

BC \ A	0	1
0 0	1	1
0 1	1	1
1 1	1	1
1 0		1

8.14 (a)

CD \ AB	0 0	0 1	1 1	1 0
0 0			1	1
0 1	1	1	1	1
1 1	1	1	1	1
1 0			1	1

(c)

CD \ AB	0 0	0 1	1 1	1 0
0 0	1	1		
0 1	1	1		
1 1	1	1		1
1 0	1	1	1	1

(e)

CD \ AB	0 0	0 1	1 1	1 0
0 0				
0 1		1		1
1 1				
1 0			1	

8.15 (d)

K-map for A = 0:

DE \ BC	00	01	11	10
0 0		1	1	
0 1	1	1	1	1
1 1	1	1	1	1
1 0				

K-map for A = 1:

DE \ BC	00	01	11	10
0 0		1	1	
0 1	1	1	1	1
1 1	1	1	1	1
1 0		1	1	

8.16 (a) $f = AC' + A'C + A'B$ (c) $f = A'B'C' + AC + BC$

8.17 (a) $f = BD + B'D'$ (c) $f = D + A'B' + B'C'$ (e) $f = BC'$ $+ AB'D' + A'CD + A'B'C$ (g) $f = A'B'CD' + AC' + BC'D'$ $+ AB'D$ (i) $f = AB'D' + ACD + A'BC + B'C'D'$

8.18 (a) $f' = BD' + B'D$ (c) $f' = BD' + ACD'$ (e) $f' = A'B'C'$ $+ BCD' + ACD + AB'D$ (g) $f' = A'D + BC + A'B'C' + ACD'$ (i) $f' = A'C + A'B'D + BCD' + ABC' + AC'D$

8.19 (a)

BC \ A	0	1
0 0		1
0 1		1
1 1	1	1
1 0	1	1

$f = A + B$

(c)

BC \ A	0	1
0 0		
0 1		1
1 1		1
1 0	1	1

$f = AC + BC'$

8.20 (a)

CD \ AB	00	01	11	10
0 0	1	1	1	1
0 1			1	1
1 1	1	1	1	1
1 0	1	1	1	1

$f = D' + C + A$

(c)

CD \ AB	0 0	0 1	1 1	1 0
0 0	1	1	1	1
0 1	1	1	1	1
1 1		1		
1 0		1		

$f = C' + A'B$

8.21 (a) $f = A + A'BC + B'C$

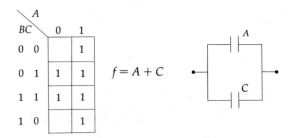

BC \ A	0	1
0 0		1
0 1	1	1
1 1	1	1
1 0		1

$f = A + C$

(c) $f = A[BC(A'B'C' + D + AD') + A'C'] = ABC(A'B'C' + D + AD') + 0 = ABCD + ABCD' = ABC$

8.22 (a) Obvious identity.

(c)

A	B	C	B $\oplus$ C	A + (B $\oplus$ C)	A + B	A + C	(A + B) $\oplus$ (A + C)
1	1	1	0	1	1	1	0
1	1	0	1	1	1	1	0
1	0	1	1	1	1	1	0
1	0	0	0	1	1	1	0
0	0	0	0	0	0	0	0
0	0	1	1	1	0	1	1
0	1	0	1	1	1	0	1
0	1	1	0	0	1	1	0

↑ Not identical   ↑ Not an identity

8.24 (a) $A' + BC = A' + (B' + C')'$.

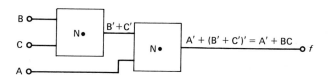

(c) Assuming 1 is available as input:

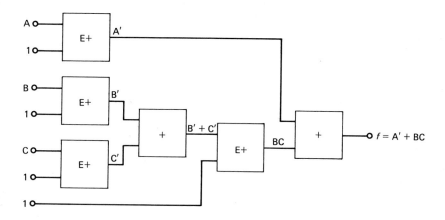

8.25 (a) Clearly the output is 1 only if $A = 1$ or $B = 1$ or both A and B are 1. This is equivalent to $A + B$.   (c) $f = AB = (A' + B')'$

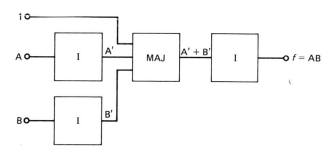

8.26 $f = A'B + CD + AB'C$

(a)

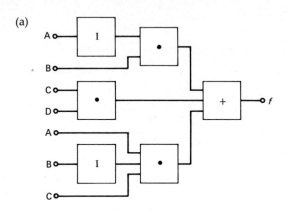

(c)

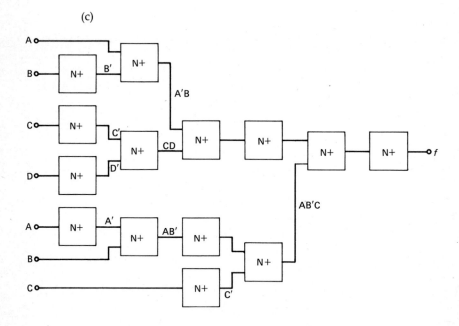

8.27 (a)

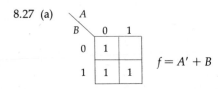

$f = A' + B$

(c) $f = A' + BC$

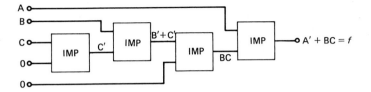

Index